KIWI FALCONS

Hamilton Asphalts
Nautech Electronics
Smart, flexible electronic solutions
30
chillex
Racing TEAM
FALCON

KIWI FALCONS

STEVE HOLMES

BATEMAN BOOKS

INTRODUCTION

Ford Australia introduced the Falcon in 1960 as an Australian-built competitor to General Motors-Holden, who'd been manufacturing cars in Australia, specifically for Australian consumers and Australian conditions, since the late 1940s.

The Australian Falcon remained in production until 2016, making it one of the longest-running name plates in world motoring history. Indeed, the original American Ford Falcon, first introduced in 1959, was quietly phased out in 1970. Its final iteration was an inconspicuous poverty-pack variant of the mid-sized Torino.

The American Falcon was created in response to an evolving market in the late 1950s. With the trend moving increasingly towards larger and more flamboyant cars, many households began shopping for a second vehicle. By and large, it was female buyers who were pushing for this second vehicle, and they wanted something smaller and more manoeuvrable than the massive land yachts that had become so prevalent. But while there was a growing selection of compacts being offered by foreign manufacturers, buyers still preferred American made. Indeed, Ford, General Motors, Chrysler, and American Motor Corporation, all entered the compact-car market in the late 1950s.

Despite intense competition in a hotly contested economy car market, Ford sold over 435,000 Falcons in 1960, and that figure increased for 1961.

The Falcon was created in the United States to perform the role of a down-beat grocery-getter, suburban runabout, and second car. But its Australian counterpart had an altogether different purpose.

In 1948, General Motors-Holden came to market with a rugged new vehicle called the Holden 48-215. The 48-215 was a rejected Chevrolet proposal that Holden management spotted during a visit to Detroit. They could see it being adapted to suit the unique Australian climate and broad variety of roads, which ranged from paved through to gravel and dirt.

Prior to the 48-215, GM-Holden produced vehicle bodies for imported chassis, including those of international General Motors brands Chevrolet, Buick, Pontiac, and Vauxhall.

Ford Australia, meanwhile, assembled completely knocked down (CKD) kits of American and Canadian vehicles before the Second World War, and predominantly British vehicles after it. But during the 1950s, and with its British Zephyr encumbered by import taxes, Ford struggled to compete with the Holden. While every inch the rival on paper – and then some – the Zephyr, plus its Zodiac and Consul derivatives, was vastly more expensive. What Ford needed to do was to follow Holden's lead and produce its own Australian car.

A programme was launched in terms of which the expensive dies required to produce the Zephyr in Australia would be purchased from Ford Great Britain, thus side-stepping the crippling import taxes. But those plans were promptly cancelled when, during a visit to Ford headquarters in Detroit during 1958, Ford Australia management viewed the new American Falcon for the first time, and recognised instantly its suitability for the Australian market. By now, the 48-215 (unofficially dubbed the FX) had been replaced by an updated variant: the Holden FJ, which itself had been succeeded by the new FE. But the Falcon was superior in almost every area – it was longer, lower, and significantly wider. It could seat six people in relative comfort. Its styling was sleeker; it looked like a 1960s car, whereas the FE was mired in the mid-1950s. And the Falcon also sported a larger, more powerful engine.

The 1960 Australian Falcon was effectively a right-hand drive variant of its American cousin. Whereas the US version offered both two-door and four-door sedan and station wagon body styles, plus a pick-up Ranchero, only the four-door options made it to Australia, as well as the Ranchero. These small car-based utility vehicles had enjoyed massive popularity in Australia going back prior to the War, and that trend continued with the Falcon 'ute'.

The 1960 Australian Falcon was called the XK. It was launched in September and remained in production until superseded by the XL in 1962.

The XL was essentially an evolution of the XK, sporting various subtle visual and mechanical upgrades. Its aesthetics were still very similar to those of its American counterpart, including its new slimmer roofline, styled on that of the Thunderbird, Ford's halo car.

In 1964 came the XM. This was the first Falcon styled in Australia, although it was clearly still heavily influenced by its American counterpart. Ford gave Falcon buyers a sporty new option by offering the two-door hardtop model for the first time. The following year saw the release of the XP, the last of the first-generation Falcons, which further built on its Australian identity.

In late 1966, Ford Australia introduced the second-generation XR Falcon. While stylistically completely different from the first-generation models, it was still, in essence, an American car built in Australia. Indeed, other than a handful of subtle styling tweaks, it was truly a beefed-up version of the third-generation American model, which had been launched the previous year. The XR saw the Australian Falcon equipped with a V8 engine for the first time; something American customers had enjoyed since 1963. However, the two-door hardtop body shape was dropped, even though it remained as part of the American range.

As the Falcon grew ever-more popular in Australia, so its presence in the US diminished, and when the American Falcon was phased out following the 1970 model year, the Australians had little choice but to go it alone. The second-generation Falcon evolved from the XR, to the XT in 1968, the XW in 1969, and finally, the XY in late 1970.

By the time the new third-generation XA was launched in 1972, there was no longer an American Falcon. The XA, therefore, became a true Australian car. Its styling was still heavily influenced by Ford Detroit DNA, in this case, the mid-sized Torino, but now the XA Falcon

could never be mistaken for anything other than Australian. To counter the sporty-looking Monaro two-door coupé offered by Holden in its HK – HG range, and its new-for-1971 HQ, Ford also reintroduced its own two-door hardtop coupé. The third-generation Falcon evolved through the XB in 1973 and the XC in 1976.

By the late 1970s, American styling influences were losing favour, replaced by those from Europe. Round edges and large front and rear overhang made way for crisp lines, large greenhouse areas, thin window pillars, and squared edges. The fourth-generation Falcon was called the XD. It was released in 1979, and looked nothing like anything that had emerged from the US. It did, however, closely resemble the second-generation European Ford Granada. Indeed, this was the direction in which Australian manufacturers were now heading. Holden's latest offering, the very first Commodore, hit the market in 1978, and on the surface it was virtually indistinguishable from the Vauxhall Viceroy/Opel Commodore C, other than its grille treatment. Of course, both the Falcon and Commodore offered a V8 in their line-ups; something unimaginable in a sophisticated Euro-box. But the brawny V8 was now very much an integral part of the Australian motoring landscape.

The fourth-generation Falcon, which spanned the models XD, XE, and XF, underwent numerous highs and lows. The late 1970s were a time of uncertainty in the automotive world, with the energy crisis still casting a shadow. Manufacturers were tending towards smaller, more economical cars, and Ford's Australian rival, Holden, followed suit with its new Commodore. The XD Falcon was notably larger than the VB Commodore. Initially, it appeared Holden had chosen the right path, as the Commodore outsold the Falcon.

1982 was an interesting year for the Falcon. Ford Australia had spearheaded the down-under muscle car craze in the late 1960s, starting with the XR in 1966, and pushed the envelope further with each new model, reaching the pinnacle with the GTHO Phase III in 1971. But then followed a gradual decline in both its motor-racing involvement and in its production of high-horsepower sporty sedans. Ford hit its lowest ebb in November 1982 when it ceased production of V8-powered Falcons. The company had completely turned its back on performance cars.

Demand for V8 cars in Australia was a fraction of overall sales. The bent-8 was purely a marketing exercise. And yet, the Falcon finally overtook its Holden rival in 1982, topping it as the best-selling car, with more than 84,000 units sold, versus 78,429 for the Commodore.

European design once again helped fashion the fifth-generation Falcon – the 1988 EA. Again, it was the Granada/Scorpio. But unlike the XD, the EA had very much carved its own identity. It was also the first Falcon to employ wind-tunnel testing to help mould its shape. The facelifted EB followed in 1991, and with it, the return of a V8 engine option. The EF, released in 1994, was the fourth iteration of the fifth-generation Falcon, and it underwent dramatic styling changes that fully embraced the popular mid-1990s jelly-mould trend, where anything resembling a hard edge was rounded off.

In 1998, Ford Australia launched the sixth-generation Falcon, the AU. The AU dove head-first into Ford's self-titled New Edge vehicle design, which influenced the styling of its world cars during this era, including the tiny Ka and the Focus. New Edge was designed to appeal to younger buyers, but it polarised opinion. And this was very much the case with the AU, which the more conservative Australian market wasn't quite ready for. Ford considered various options to replace the outgoing fifth-generation Falcon, including introducing the American Taurus, or the European Granada. In the end, it retained the fifth-generation architecture, mated with a radically redesigned shape and numerous mechanical upgrades.

But Ford had to quickly back-pedal when the response to the new AU was less than flattering. The AUII (launched in 2000) and the AUIII, a year later, addressed some of the criticism. But it was the BA, released in late 2002, that finally wiped the slate clean. The BA was a heavily made-over AU, sporting a longer wheelbase, and far more conventional styling. Of the exterior panels, only the door skins were carried across. To distance itself from the AU, it was assigned a new designation, and the BA prompted a comeback for the Falcon.

The mid-2000s saw record new car sales in Australia, but by the time the seventh-generation Falcon, the FG, was launched in 2008, there were signs the market was beginning to change. Having topped the charts for well over a decade, the Commodore was knocked off the top spot in 2011 by the Mazda 3, and there began a gradual decline in sales for the big domestic sedans.

In 2014, Ford released the FG-X, a facelifted and improved version of the FG/FG II. Upon its launch, it was already common knowledge that this would be the very last Falcon. By and large, this was the best car Ford Australia had ever produced, and it garnered praise from the motoring media. Its styling adopted the Aston Martin-style nose incorporated into many of Ford's international models, including the Mustang, Mondeo, and others.

But customers didn't want full-size sedans anymore. They'd changed their allegiance to small compacts, SUV's, and large pick-ups. There was truly no coming back for the Falcon. Unlike Holden, who attempted to continue selling the Commodore using a rebadged Opel Insignia in 2018, Ford Australia instead retired the Falcon, and the last model rolled off the Broadmeadows assembly line in 2016. And thus ended 56 years of production, and the longest-running name plate in the international history of Ford Motor Company.

The Falcon in Racing

An important chapter in Falcon history is its racing pedigree. Both the American and Australian Falcon featured in factory motor-racing programmes. The US variant took part in European rallying during 1963 and 1964. When Ford launched the new Falcon Sprint with its small block V8 engine in 1963, it embarked on a racing campaign that became just a small cog in a much larger motorsport programme taking place at the time.

In the early 1960s, as part of its ambitious Total Performance marketing program, Ford US took the Falcon racing. The company was already competing at the highest level in NASCAR stock car racing and NHRA drag racing, and began a colossal project to beat Ferrari at Le Mans, beginning with the 1964 French epic. The Ford Galaxie and Mercury Marauder were competing in NASCAR, while the Galaxie, Fairlane, and Mercury Comet were all represented in NHRA.

Ford wanted to introduce the Falcon into its performance programme and improve its dowdy image – a result of its original designation as a cheap and cheerful economy car. In 1963, the Sprint was launched as part of the Falcon range. It sported a handsome two-door hardtop body style, and was optioned with a 260 cu inch V8 engine. To help thrust the Falcon into the thought processes of new-car buyers, Ford had Holman Moody, who ran one of its stock car teams, build a fleet of Falcon Sprints to contest European rallies. Alan Mann Racing, the Ford of Britain factory team, then ran the cars in a series of rally events, including the Monte Carlo and Tour de France. The success of this programme ensured that it continued into 1964 before the new Mustang took its place late in the year.

But just as the US Falcon was being retrofitted to its original role as an unassuming grocery-getter, Ford Australia was beginning to ramp up the Falcon's competition involvement down-under.

The Falcon had been competing in racing events in Australia as early as 1960. Melbourne car dealer Bob Jane and Lou Molina shared a Falcon XK at the 1960 Armstrong 500. This was a 500-mile (805-kilometre) endurance race for Australian-made or Australian-assembled cars, contested at the Phillip Island track in Victoria. A second Falcon was shared by Ron Phillips and Ern Seeliger. Vehicles were classed based on engine size, and were required to be completely stock-standard other than having items such as hubcaps and other objects that could pose a safety hazard removed. The Jane/Molina Falcon finished third in Class D, behind a Vauxhall Cresta and a Standard Vanguard. Outright positions weren't recognised in the race; only class positions.

Only one Falcon contested the 1961 Armstrong 500. It was shared by Ken Harper, Syd Fisher, and John Raeburn, and it finished second in Class C, albeit six laps down on the class-winning Mercedes-Benz 220SE of Bob Jane and Harry Firth.

Ford Australia decided to enter the fray in an official capacity for the 1962 race, supporting an official two-car team of Falcon XKs, driven by Jane/Firth, and Harper/Fisher/Raeburn, while a further three Falcons were also entered by privateer teams. After 500 miles, the Jane/Firth Falcon crossed the line first.

During the 1962 race, the track surface broke apart, causing damage to several cars, and without the funding needed to make repairs, the Armstrong 500 was moved to a different facility for 1963. The new venue was an arduous 6.13-kilometre-long public road circuit near the small New South Wales town of Bathurst. The road snaked its way up and across the top of Mount Panorama. From its lowest point at the start line, to its highest at the top of the mountain, there was a vertical difference of 174 metres. While there were countless permanent race tracks around Australia, Mount Panorama was chosen specifically because it tested the capabilities and durability of the vehicles competing, much more so than any other track. To win was the ultimate reward. But just to finish was itself an accomplishment.

The regulations stated that in order to be eligible, 100 cars had to be built in Australia, or 250 had to be imported/assembled. To that end, Holden spotted an opportunity to gain some healthy publicity, and built a special variant of its new EH, specifically for the Bathurst race. It was called the EH 225 M-S4, and six examples entered the 1963 event. The S4 was, however, a slight disappointment, being beaten soundly by Ford's Cortina GT. For the 1963 Armstrong 500, Ford Australia dropped the Falcon in favour of the Cortina. Four GTs were entered in Class C, and they finished first, third and fourth, split only by the leading S4. It was the magic pairing of Bob Jane and Harry Firth who again shone. Holden quietly withdrew the S4 and the 1964 Bathurst 500 was a repeat of the 1963 event, only this time even more emphatic was the victory, as Cortina GTs filled the top six finishing positions, with Bob Jane this time sharing the winning car with George Reynolds.

For 1965, Ford Australia followed Holden's lead, and produced a Bathurst special. It was called the Cortina GT500, and Barry Seton/Midge Bosworth drove one of these to victory. 1965 was the first year in which the outright winner was recognised.

For 1966, the Australian Racing Drivers' Club, which ran the Bathurst event, opted to increase the minimum number of cars required to contest the Bathurst enduro, from 100 locally produced cars, to 250. This eliminated the Cortina GT500, and an army of Mini Cooper S's stole the show, filling the first nine positions at the finish.

For 1967, however, Ford Australia returned. Its latest weapon was a

XL Falcons were thin on the ground in New Zealand in the 1960s, and even more so today. They're now highly collectable.

V8-powered version of its new Falcon XR model, specifically designed to win at Bathurst. It was called the Falcon GT. It was fitted with an uprated version of the 289 cu inch V8, producing 225 horsepower, plus a range of performance hardware throughout the driveline, suspension and brakes. All 558 XR Falcon GTs were painted gold, with the exception of two special-order examples in white. Remarkably, this was the very first time the Ford Falcon was represented in the Bathurst endurance race. And it would win on debut. Seven of the new XR Falcon GTs were entered, including three works entries. Two of the factory examples, driven by Harry Firth/Fred Gibson, and brothers Ian and Leo Geoghegan placed first and second. It was an impressive showing.

For 1968, Ford released the next version of the Falcon GT, based on the latest XT model. Ford improved the car in several areas, including the engine, which was now at 302 cu inch and producing 230 horsepower. The XT was a better all-round car than the XR, but it wasn't good enough to win Bathurst. Holden returned in 1968, with its latest sporty offering, the HK Monaro GTS327. Its 327 cu inch Chevrolet V8 produced 250 horsepower, and it was impressively quick. Seven Monaros were entered; three claimed the top three fastest qualifying times, and Monaros also swept the podium positions. Although there were three factory Monaros for the Holden Dealer Team, it was the privateer entry of Bruce McPhee and Barry Mulholland that took pole and the win.

Holden had come at Ford with intent, and Ford responded in 1969 with its latest Bathurst warrior, based on the facelifted XW Falcon. Although Ford still produced the Falcon GT, for 1969, it came out swinging with an even more potent model, called the GTHO. From a total 2949 Falcon GTs produced in 1969, only 662 were the GTHO. This was a car designed specifically to win Bathurst. Engine size and horsepower made another big leap over the previous year, with a 351 cu inch V8 producing 300 horsepower. Meanwhile, Holden returned with its new HT Monaro GTS350, which, as its name suggests, was powered by a larger 350 cu inch Chevrolet engine. Like the GTHO, it packed 300 horsepower.

Bobby Owens' beautiful 1964 XM Wild Bunch Falcon hardtop drag car. It runs a blown 509 cu inch big block and regularly punches out low 7-second quarter-mile times.

Fourteen XW Falcon GTHOs took the starter's flag at Bathurst, including three Ford Australia factory entries. They faced off against seven HT Monaro GTS350s, including three Holden Dealer Team entries. It was Ian Geoghegan who took pole in his GTHO, followed by Digby Cooke's GTS350. Five more GTHOs and three GTS350s locked out the top ten qualifying spots. However, the race results were turned on their head when, on the opening lap, Bill Brown inverted his GTHO at the top of the mountain, with several of the front-runners getting collected in the drama. Throughout the race, even though they were faster, several of the remaining Falcons struck tyre problems, and when the chequered flag fell, it was the Holden Dealer Team Monaro, driven by Colin Bond and Tony Roberts, that was ahead. Bruce McPhee and Barry Mulholland, who'd switched from Holden to Ford for 1969, placed second in their privateer GTHO, while Peter Brock, making his Bathurst debut, finished third, in the HDT Monaro he shared with Des West. Behind them, in the first of the Ford Australia factory GHTOs, was another driver making his Bathurst debut. Allan Moffat appeared likely to win Bathurst at his first attempt, but was called in by his team for tyres late in the race. All the other Falcons were destroying their tyres, and the team felt sure the same would be the case for Moffat. But somehow he was not only conserving his tyres, he was doing so while producing record-setting laps. Moffat's stop almost certainly cost him the race win on debut.

For 1970, however, Moffat would exact his revenge. Ford's latest Bathurst special was the XW GTHO Phase II, a better and faster version of the previous year's model. Although still an XW model, and outwardly very similar to its predecessor, the Phase II sported a 351 cu inch V8 using one of the new Ford Cleveland engine blocks, replacing the outgoing Windsor of the previous models. Meanwhile, Holden opted to take another route in its bid for Bathurst glory. It dropped the V8 Monaro in favour of the small Holden Torana, which was powered by a 186 cu inch inline-6. Holden felt a smaller, more nimble car that better preserved its brakes and tyres would prove a better weapon of choice. In 1970, at least, it was wrong. Seven of the top ten cars in qualifying were GTHO Phase II Falcons, including the first three. And in the race, while the little Holdens did indeed look after their brakes and tyres better than the fire-breathing Fords, it

Awesome (and rare!) Falcon XP wagon, owned by father and son collectors Ross and Dallas.

wasn't enough. Moffat took his first Bathurst victory, ahead of Bruce McPhee, both driving works Falcons. The first Torana home was that of Don Holland, who was a lap behind Moffat at the end.

The final iteration of the GTHO Falcon was launched in September, 1971. The Phase III was based on the facelifted XY model, and it was a faster, more powerful beast again. Its most prominent feature was a centre 'shaker' scoop mounted on top of the carburettor, poking up through a hole in the bonnet. The Phase III possessed around 380 horsepower, although Ford listed its power at a more conservative 300. It was still fitted with tiny 14-inch-diameter wheels, but was considered to be the fastest accelerating four-door sedan on the planet.

Although Holden released a more powerful variant of its Torana XU1, with 202 cubic inches, the big Fords were even more dominant at the mountain than they had been the year before. The first seven cars in qualifying were all GTHO Phase IIIs. No Toranas even qualified in the top ten. The nearest anyone could get to Ford's muscle car was the handsome new Valiant Charger E38. Leo Geoghegan qualified his factory example in eighth. Come the race, and while the little Toranas moved forward as they better conserved their components and could run longer stints between pit stops, in the end, three Phase IIIs clean-swept the podium headed by Moffat.

Ford was in the process of building its next version of the GTHO, the Phase IV, in 1972. This was to be based on the new XA-model Falcon. However, a story by motoring writer Evan Green in the *Sydney Morning Herald* sensationalising the deadly 'Supercars' being produced by Ford, Holden, and Chrysler Australia sparked a flurry of negative media attention, and prompted all three manufacturers to promptly cancel their upcoming plans. Rather than racing the XE GTHO Phase IV at Bathurst in 1972, Ford instead reverted back to the 1971 Phase III.

The XA did, however, get to compete at the highest level, beginning in 1973. Following the Supercar scare, the Australian motorsport governing body quickly sidelined the Series Production regulations, and replaced them with Group C. Series production required the cars to be completely stock standard – just as they rolled off the assembly line. To achieve Bathurst success with such limitation meant building a hugely potent production vehicle. Group C side-stepped this issue by allowing numerous upgrades to the standard vehicle, so that it would not be required to be so potent in standard form.

Group C also became the new set of regulations under which the Australian Touring Car Championship (ATCC) was contested. Up until 1972, the ATCC used Improved Production rules, which favoured the fast American pony cars: the Mustangs and Camaros. Ford Australia funded a programme to build a pair of highly powerful Improved Production XW (updated to XY) Falcons in 1970, for Ford factory drivers Allan Moffat and Ian 'Pete' Geoghegan. Moffat didn't much care for his Super Falcon, and reverted back to his much-loved 1969 Mustang. Geoghegan, however, persevered, and continued development of his example throughout 1971 and '72. Although still not a match for the American cars, Geoghegan did take the very first ATCC race victory for the Falcon model at the Mount Panorama round of the championship.

For 1973, Allan Moffat won the ATCC in his upgraded Series Production GTHO Phase III, before switching to a newly constructed XA GT hardtop to win the Bathurst enduro. Moffat also won the ATCC in 1976 and '77 (in a Falcon XB GT), and won Bathurst in 1977, with the latest XC model. John Goss, driving an XA GT, won Bathurst in 1974.

The Group C regulations were given a shake-up for 1980, to allow the new Falcon XD and the Commodore to compete. A new Ford hero emerged in the form of Queenslander, Dick Johnson. Johnson won the crash-shortened 1981 Bathurst race, and also won the ATCC in 1981, 1982, and 1984. The 1984 championship was achieved aboard an XE, even though Ford had stopped production of V8 Falcons two years earlier.

International Group A regulations replaced Group A for 1985, and with the Falcon not homologated, Ford teams competed with the third-generation Mustang, until the wicked new Sierra Cosworth/RS500 was ready in 1987. Indeed, the Falcon didn't return to front-line Australian touring car racing until the advent of the new 5-Litre Touring Car formula was launched in 1993. This new platform pitted the Falcon against the Commodore, in a special Australian-made concoction that would be re-branded as V8 Supercars and, eventually, Supercars. Mercedes-Benz, Nissan and Volvo also competed but rarely challenged the 'Big Two'.

Glenn Seton, John Bowe, Marcos Ambrose, Russell Ingall, Jamie Whincup, James Courtney, Mark Winterbottom, and Scott McLaughlin all won Supercars Championships with the Falcon. Indeed, McLaughlin's 2018 title was achieved in the last year the Falcon was eligible to compete, having been out of production for two years.

Likewise, Dick Johnson/John Bowe, Jason Bright/Steven Richards, Craig Lowndes/Jamie Whincup, Mark Winterbottom/Steven Richards, and Chaz Mostert/Paul Morris, all scored Bathurst 1000 victories for the Falcon in the Supercars era.

The Falcon also has its place in New Zealand motor racing. The model was hugely popular, and highly successful in the competitive Castrol GTX production car series of the 1970s. Jim Richards won the 1972 championship driving a Falcon GTHO Phase II, and again in 1974 and '75, with an XA GT sedan. Barry Lloyd won the championship in 1978, driving an XA GT hardtop. This was the final year the championship was contested.

Production car racing was enormously popular in New Zealand from the 1960s through 1980s, thanks to the strong domestic assembly industry. The Wills 6-Hour, which evolved into the Benson & Hedges 500, was an annual endurance event for New Zealand-assembled vehicles. It was dominated throughout the 1970s by the Valiant Charger 770, but when the race was redesigned as an endurance series in the 1980s, Leo Leonard and Gary Sprague won the championship driving a Fairmont Ghia. In the early 1980s, the Fairmont, with its 302 cu inch V8 and automatic transmission, was the most potent locally assembled Ford on the market.

Following a change to the regulations, Gary Sprague won the 1984 New Zealand Touring Car Championship, driving a 351 cu inch Falcon XD.

Just as had happened in Australia, New Zealand adopted the international Group A regulations beginning with the 1984 championship. Group A was also used for the very first Wellington Street Race, contested in January 1985. Driving a 351 cu inch XE Falcon owned by Bruce and Wayne Anderson, Dick Johnson teamed up with Bruce and took pole position, beating off some of the fastest Group A machines on the planet.

In 1996, piggybacking off the success of the Australian V8 Supercars Championship, New Zealand created the NZV8 Touring Car Championship. It incorporated a set of locally produced regulations to restrict costs, but allowing the cars themselves to closely resemble the Australian cars. As with the Australian series, NZV8s only allowed Holden Commodores and Ford Falcons to compete. Ashley Stichbury, Mark Pedersen, John McIntyre, Kayne Scott, Craig Baird, and Angus Fogg all won the championship in Falcons before the series underwent a regulation change.

Beyond the front-line combat, Falcon has always been a popular model of choice among Kiwi racers, particularly in muscle car and historic car racing. Its growing list of enthusiast owners have taken part in everything from Dunlop Targa, through to muscle car racing and historic racing.

The Falcon in New Zealand

The Falcon wasn't purely an Australian icon. It was also embraced into Kiwi culture. When Ford Australia launched the Falcon in 1960, Ford New Zealand continued selling the Zephyr. Although a handful of Falcons did trickle across the pond in the early 1960s, it wasn't until the third iteration of the first-generation, the XM, that Falcon officially arrived. But unlike Australia where Zephyr supply was switched off with the arrival of the Falcon, in New Zealand, customers had a choice between the British and Aussie big Fords. The cross-over included both the MkIII Zephyr and the MkIV. But by the early 1970s, only the Falcon remained.

Falcons were assembled in New Zealand from CKD packs, while the more specialised models, such as the GTs, were imported as complete cars. This remained the case for decades, until eventually they were only imported complete.

For decades, the full-sized Australian Ford and Holden sedans appeared the perfect vehicles for the New Zealand environment. Cheap petrol and rough roads required a specific vehicle type, and these cars fit the bill. They served well towing the caravan, or a trailer. They effectively performed the duty of a work vehicle and comfortable family car. They were available in multiple body styles to suit every need. They were affordable to purchase, and to repair and maintain. Bang-for-buck, nothing else came close. But by the late 2000s, as was happening in Australia, buyer patterns began to change. Suddenly, people started turning towards smaller cars, or SUVs, or large utility vehicles. Indeed, these utility vehicles that had once been basic and rudimentary were becoming increasingly comfortable; able to travel long distances with composure, while also performing heavy work duties. They sported all the comforts and conveniences of any modern car.

With so many choices on offer, demand for large family sedans slumped. Indeed, in 2016 – the year Ford Australia ended Falcon production – only one 'car', the Toyota Corolla, made the top ten best selling vehicles in New Zealand. The rest were utility vehicles, headed by the Ford Ranger, or SUVs. No amount of history could save the Falcon.

Ford may have stopped building Falcons, but as long as oil is being extracted to keep them running, owners will continue enjoying them with passion and pride. The Falcon has a following in New Zealand that will never fade, and in the years to come, as numbers of current models are struck down, they will even gain in desirability. Although this was not apparent at the time, when the very last example rolled off the Broadmeadows assembly line in 2016, every Ford Falcon instantly became a future classic.

FALCON

BRIAN'S

XK FALCON

Some significant Ford Falcons have been built over the years, but none more significant than the XK.

During the late 1950s, Holden was enjoying runaway sales success with its succession of Australian-built vehicles, designed specifically for local conditions. Beginning with the 48-215 in 1948, Holden built cars that were designed to cope with rugged Australian roads, that carried a family and their luggage in relative comfort, and that was priced to meet the budgets of Australian buyers. This was a unique car for unique conditions, and subsequent Holdens followed suit.

Ford Australia, however, struggled to compete. It was still assembling the British Zephyr and Consul models, and while these were all-round good cars, because Ford was lumbered with import tariffs, it struggled to match Holden's pricing. And as a result, Zephyr and Consul sales also struggled.

Ford's only option was to manufacturer its own car, to compete head-to-head with Holden. Initially it intended to produce the Consul, Zephyr, and Zodiac locally, but would need to purchase the expensive dies from Ford Britain. However, it was during a visit to Ford's Detroit headquarters in 1958 that Australian executives were shown the company's upcoming new compact car, the Falcon. Immediately they were sold on the concept.

The Falcon was a project being pushed along by Ford General Manager Robert S. McNamara. During the late 1950s, Detroit's three big car manufacturers -Ford, General Motors, and Chrysler – were all

developing small cars in response to market demands. Throughout the 1950s, cars had become increasingly large and expensive to produce. But the market was changing. Research showed that many families were looking to purchase a second car, and that there was an increase in women buyers who disliked the sheer scale of the full-size models.

The Falcon was designed as cheap, basic transport. By worldwide standards, the Falcon was a mid-size car, but by US standards, it was a compact. For Ford Australia, however, it offered the perfect solution to counter Holden's runaway success. By the late 1950s, Holden was selling its FC model, which would be replaced in 1960 by the FB. By comparison, the new Falcon was a more modern-looking car and could seat six people, yet was a full 100 kilograms lighter than a Zephyr, so it was also economical.

In 1959, a new factory was constructed in Campbellfield, Melbourne, in preparation for Australian manufacturing of a right-hand-drive version of the Falcon. The Australian offering was mostly unchanged from that being produced in the US, other than having a final drive ratio better suited to Australian conditions. The new Falcon, which followed in the Australian tradition with each model named using two letters, was dubbed the XK, and went on sale in September 1960 in four-door sedan form, with a station wagon added two months later. The station wagon lacked the rear overhang of its American cousin, as there were fears the rear might scrape the ground on rough roads.

There was only one engine offered, a 144 cu inch inline 6, which was uprated to 170 cu inch late in the model's life.

Initial sales were encouraging, helped along thanks to rival Holden's new FB proving slightly disappointing, but the Falcon began having suspension problems on rough Australian roads, particularly in the Outback, and the consequent criticism damaged the new car's reputation.

In May 1961, a utility and panel van were added to the Falcon range.

In 1962, Ford released the XK's successor, the XL, which featured much-needed beefed-up suspension components, and while the improvements added greatly to the cars' durability in the unique Australian environment, the Falcon's reputation took a little time to recover.

Come the XM in 1964, which was the first Falcon with an Australian-designed body, and the model was now well on its way to success. The Falcon would continue to be manufactured in Australia, built for Australian conditions, Australian needs, and Australian budgets, until 2016. Indeed, the Australian Falcon far outlived the American variant from which it originally stemmed. The Falcon name in the US lasted only until 1970, by which time it was being offered as a bare-bones version of the Fairlane – itself a bare-bones version of the Torino.

XK Falcons are extremely rare in New Zealand. Ford New Zealand was enjoying plenty of success with the Zephyr, and in 1960 the new

0 10 20 30 40 50 60 70 80 90 100
Falcon

Australian Falcon really didn't have an obvious place in the New Zealand market. However, early Ford enthusiast Brian owns three examples, the most notable of which is a 1960 sedan, which would have been one of the very first to roll off the then new Campbellfield assembly line.

Brian's XK has been in New Zealand since 1970. It's a factory two-tone green car, and was also optioned with a few extras, including the extremely rare bonnet bird emblem, sun visor, radio, and heater. Brian purchased the car in 1995 from an owner who was looking to scrap it. The previous owner had had plans to try to shoehorn a 351 cu inch Cleveland V8 into its engine bay, but had lost interest in the project. When Brian purchased the car it still, thankfully, had its original 6-cylinder motor fitted. It didn't run at the time, but the fix proved to be as simple as dealing with a broken fuel line. Once fixed, the old car fired up straightaway.

Brian repainted the Falcon in its original two-tone green hue not long after purchasing it, but otherwise it's completely original, right down to the upholstery. His passion for early Falcons means he also owns a pair of 1961 XK station wagons and some impressive memorabilia, including a rare dealer promotional model. These were metal or plastic scale model cars that dealers kept on display, in addition to a glossy brochure, to help show potential customers the different cars they had available. Buyers were often given these models as gifts when purchasing a new car.

In total, Ford Australia produced just over 3.8 million Falcons. Brian's car is number 2580. The odometer reads 77,000 miles, and by all accounts it doesn't appear to have clocked over. Predictably, there have been invitations to display Brian's XK at various shows celebrating Ford, including 100 years of Ford, and 50 years of Falcon. On both occasions, it was the only XK in attendance.

Brian's amazing XK Falcon is an incredibly important piece of history – and not just of Falcon history, but also of Ford Australia's manufacturing history. The Falcon model became a staple for the Australian public, and its ongoing and healthy rivalry with Holden's equivalent models over the years has sparked much passion and debate.

The Ford Falcon became a legend that survived for 56 years, and it all began with this modest and unlikely little four-door sedan.

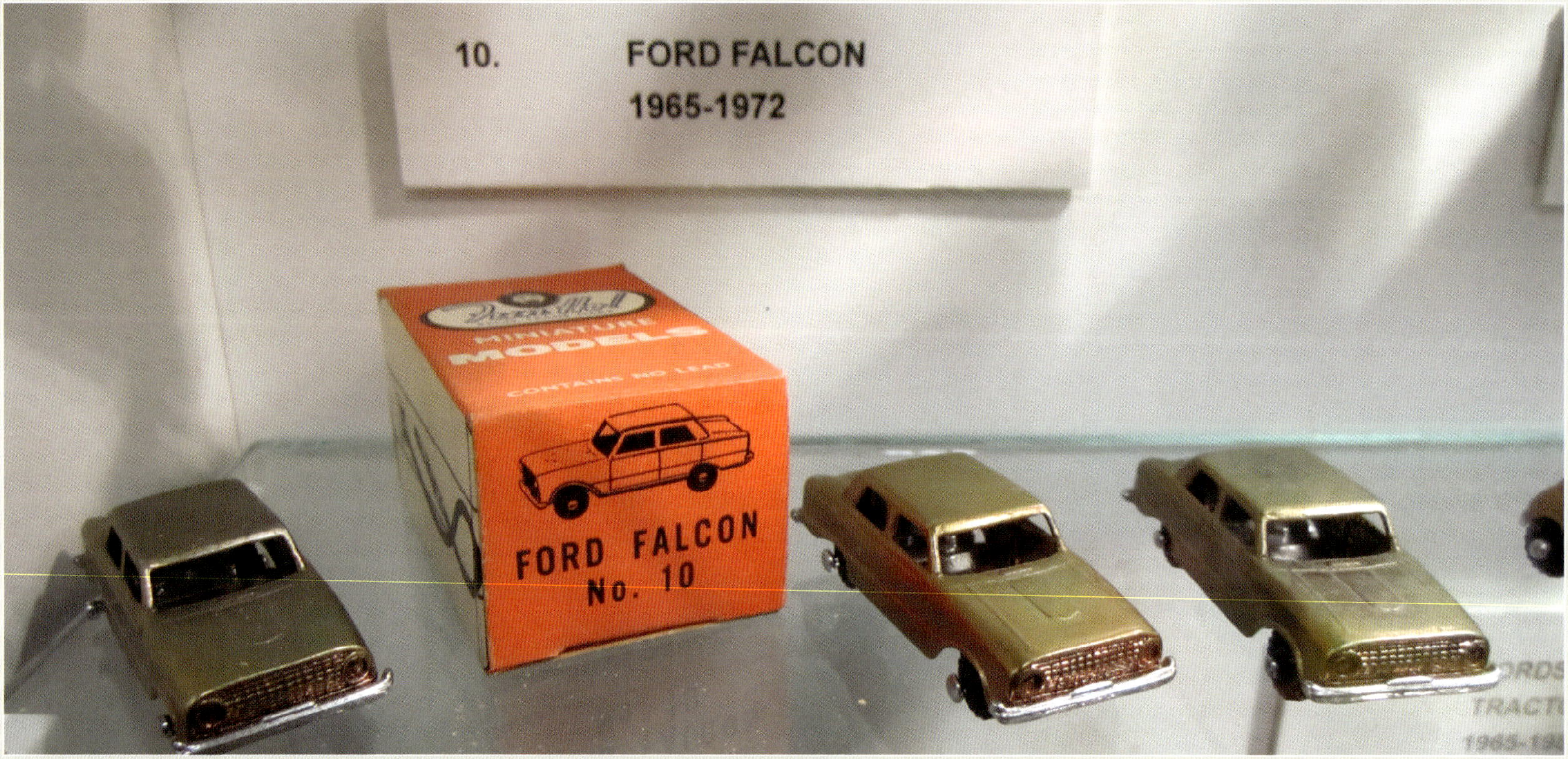

Top: Kiwi-made Falcon model cars, on display at the Fun Ho! museum in Inglewood.
Above: When Kiwi petrolhead Mike Feisst visited Australia in the late 1960s, he photographed this XR GT parked on a Sydney street. Such cars were extremely rare in New Zealand.
Right: Classic Carlectables 1:18 scale replica of the Falcon XR GT, driven by Fred Gibson and Harry Firth, which won the 1967 Bathurst 500.

Top right: XP Falcon wagon parked outside old Christchurch tenement houses in 1987.
Above: Outside Papawera Repairs & Workshop, sometime in the 1980s. The old XP was probably destined for the scrappers.

HELENA'S XP WAGON

The XP marked the fourth and final iteration of the first-generation Australian Falcon. Its lifespan, beginning in 1965, lasted just over a year before the second-generation XR was launched in September 1966. Stylistically, the XP offered the greatest breakaway yet from the American Falcon, and was the first not to feature the distinctive 'bullet' side bodywork sculpture that defined both the US and Australian first-generation cars.

The XP Falcon was available in a range of body styles, including four-door sedan, two-door hardtop, utility, two-door panel van, and this, the station wagon.

Although the wagons were conceived purely for their functionality, in recent years classic examples have gained a strong following with enthusiasts the world over. The many years during which they suffered from a lack of desirability meant their survival rates were low, which has further driven up recent demand. That they can seat more occupants than most other models makes them popular show cruisers.

Helena's 1966 XP wagon is a stunning example of everything that is great about early Falcon wagons. It has been modernised, somewhat, with the fitment of a 4.1 litre engine from an XD Falcon. It runs a manual three-on-the-tree transmission, and is lowered 2 inches in all four corners just to give it the perfect stance. The wheels have a 14 inch diameter (XP Falcons rolled off the assembly line on 13-inch wheels!). The colour looks, at first glance, to be a mid-1960s Ford factory hue, but is in fact a home-brew blue.

FALCON

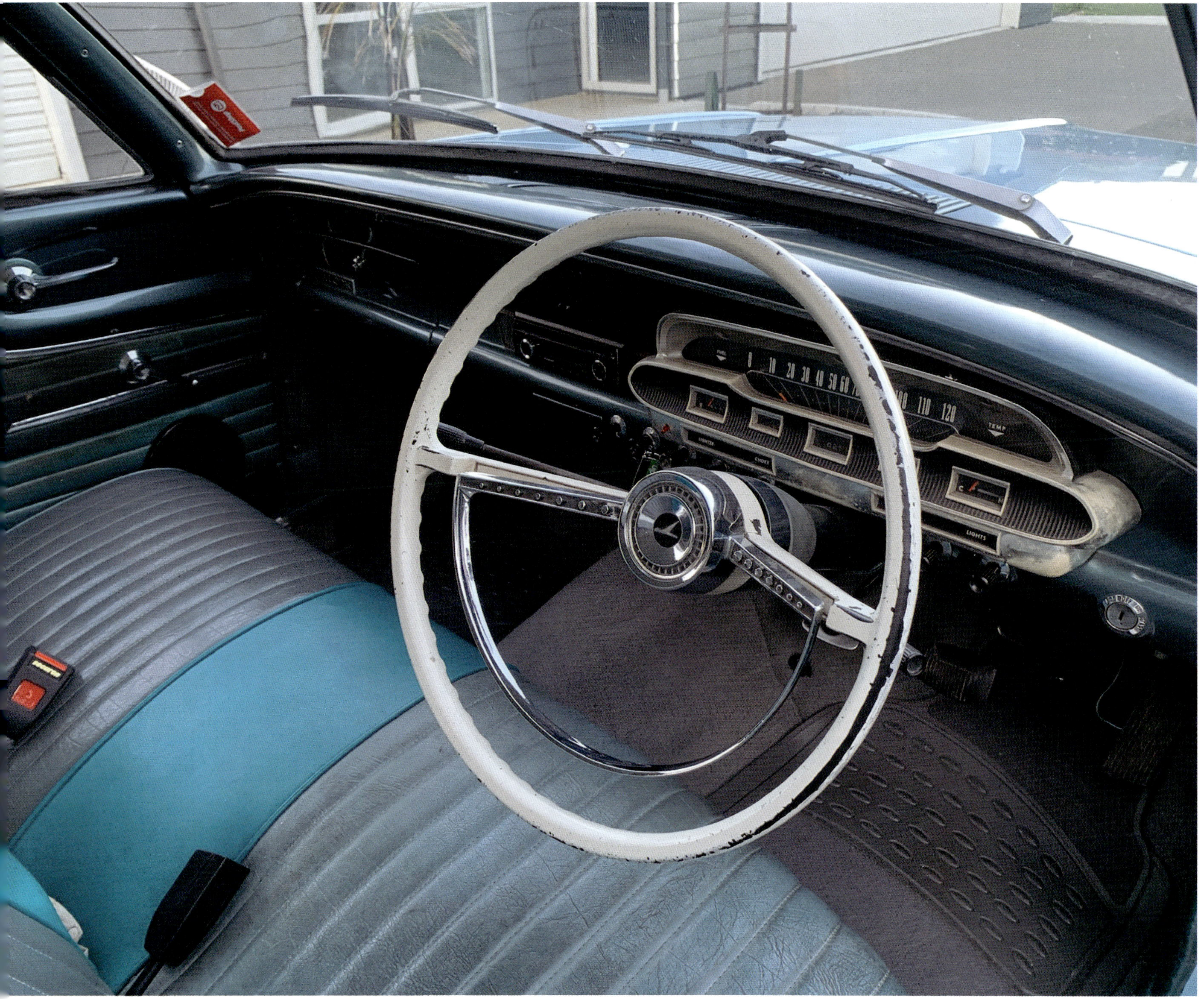
LIGHTER
CHOKE
TEMP
LIGHTS

Stewart Quertier, long-time owner of the great Ford flathead V8 powered Stuart Special, used his Falcon XT as his tow car. And sometimes the Falcon even saw a little action itself.

Classic Carlectables 1:18 scale replica of the Falcon XR GT driven to second position in the 1967 Bathurst 500 by brothers Leo and Ian Geoghegan. The brothers were sponsored by Castrol, and all their race cars were this colour scheme.

Above: Meremere (Champion Dragway) course and promotional XW Falcon wagon, hand-painted (as they all were back then) in reference to its upcoming launch in 1973.
Opposite bottom right: Legendary racer Ernie Sprague owned this Falcon XW GT in the early 1970s. It doubled as regular daily transport and circuit racer.

FORD
FALCON

500

Make up
our own
Grand Sport
rst, pick your model . . . Falcon 500 . . .
tura . . . or Fairmont Sedan. Then add
e GS Option Package, and exciting
ngs happen.
Along the side, raking GS Rally
ripes sweep to a GS Insignia. Heavily-
romed GS Full Wheel Covers add to
e sporty flavour and show your exu-
rant mood. Inside, there's a deep-
sh 3-Spoke Wood-Grained Sports
eering Wheel with a Full-Circle, Rim-
ueeze Horn and a GS Cockpit Dash
at really lets you know what's
ppening.
You take it from there. Add Bucket
eats and a Sports Console that
uses your choice of 4-speed, All-
nchro Manual Shift or T-Bar Auto-
atic. Mustang 5-litre, 220-horse-
wer V8. 11¼-inch Turbo-cooled
wer Disc Brakes. Power Steering.
r-Conditioning. Tinted and Lami-
ted Windscreen. Vinyl Roof. Wide-
al Radial Tyres that put extra tread
the road, giving your GS an even
wer sportier look.
Falcon's great list of options
etches on and on. The Falcon you
oose can be a luxury car . . . sports
r . . . family fun car . . . or a blend of
three, just the way you want it to be.
Above: GS Interior with Optional Buckets
Wheel Covers
Sports Console and Floor Shift
GS Cockpit-type Instrument Panel and Steering Wheel

351-GT
GT

*Ross Francis races this genuine Falcon XT GT in classic racing events, and has done for decades. He is part of the loosely formed AFRICA (Another F%#*ing Race I Can't Afford) Racing Team.*

The PDL Falcon, built and raced by Clyde Collins from 1969 through to 1971, before the team purchased a Mustang from Paul Fahey.

Top: Beautiful XW Falcon pictured at a car show at Ruapuna in the late 1990s.
Above: Back in the 1990s, classic Falcons were a common sight on Kiwi roads, and drew little attention. This dusty XW is pictured in Northland in 1995.

XW Falcon ute warms the rear tyres in preparation for a quarter-mile blast.

GT
FALCON

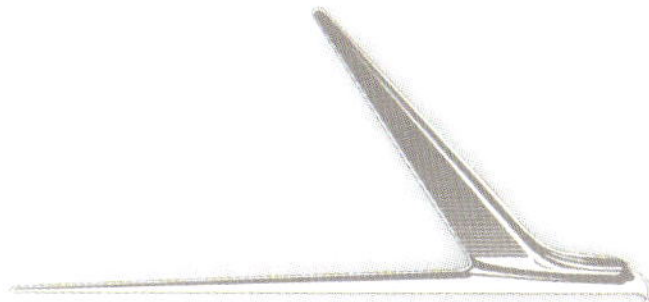

ROB'S FALCON GTHOs

Up until the end of 1972, Australia had two very different and unique touring car racing categories. Each had its own distinct set of regulations. And each had its own important crown jewel carrying such weight that it brought with it manufacturer involvement and investment. These two categories were called Improved Production and Series Production.

Of the two, Improved Production had the more liberal rules. It was first introduced in 1965, and was largely based on the FIA's Group 2 touring car regulations which were in use throughout the UK, Europe and the US, and which would eventually be adopted in New Zealand. Improved Production rules were tweaked throughout the years the formula was in place, in an effort to create some sort of level playing field.

Improved Production catered to mass-produced vehicles and homologation specials, but the way its rules were structured meant low-volume homologation specials didn't get any sort of advantage. Improved Production allowed vehicles of any origin, as long as they fitted the regulations as they were written. For this reason, the world's very best sedans took part, including those built in Australia, the US, England, and Germany. However, American cars were by far the most successful.

Improved Production was, as the name suggests, for production cars in which improvements and upgrades were allowed. Bodywork had to remain largely standard, but most areas beneath the skin could be modified to some extent. From 1970 onward, wheel widths could be as much as 10 inches, whereas prior to 1970 the regulation was 8 inches.

Fuel-injection was allowed, as were multiple carburettors, while engine sizes could be increased from the standard size. Prior to 1972, there was no maximum engine size. These freedoms, combined with the healthy budgets the top teams were armed with, meant Improved Production cars were very fast for their time.

The annual Australian Touring Car Championship, which first began in 1960 and which now runs under the guise of the V8 Supercar Championship, was contested from 1965 through to 1972 by Improved Production cars. Throughout that period, American cars won the championship every year except one (1970).

Series Production, by contrast, had an extremely tight set of regulations, in which vehicles had to be essentially as they rolled off the manufacturers' production lines. Series Production cars had to be identical to those anyone could purchase from their local dealership. Only very minor improvements could be made, such as replacing brake pads and blueprinting engines. Furthermore, the cars had to be Australian assembled or sold new in Australia.

Series Production rules required that a minimum of 5000 vehicles had to be produced in a single twelve-month period to be eligible. However, the rules also allowed for homologation specials, which were cars based on those being mass produced, but with specific upgrades of the manufacturers' choosing. But in order for these vehicles to be eligible, a minimum 100 cars had to be produced locally, or 250 imported, in a twelve-month period.

There were multiple classes taking part within the Series Production rules, and these classes were based not on engine size, but on retail price.

Unlike Improved Production there was no high-profile national championship for Series Production cars. However, once a year they contested a 500-mile (800-kilometre) endurance race that was initially held at Phillip Island racing circuit, in Victoria, from 1960 through to 1962, before it was moved to an extensive and hugely challenging road course circuit called Mount Panorama, on the outskirts of a small New South Wales town called Bathurst. And it was the annual Bathurst endurance race that became the crown jewel for Series Production cars.

Holden was the first manufacturer to recognise that achieving success at Bathurst provided excellent promotional leverage to boast about the speed, durability and quality of its motor cars. Therefore, in 1963, Holden produced a Bathurst homologation special of its EH model called the S4. The S4 featured a number of upgrades over the standard EH model, specifically aimed at helping it win the Bathurst race. It failed to achieve its objective, but the gauntlet had been thrown down.

Ford won Bathurst in both 1963 and 1964 with its Cortina model. However, Holden was planning an upgraded version of the S4 for 1964, called the S22. Ultimately, this project was stillborn, although The General did incorporate some of the features from the S22 into its latest model, the HD, for 1965. This it named the X2. Likewise, British Motor Corporation (BMC) planned to sell the new 1275 cc Mini Cooper S into Australia using Bathurst as a launching pad. So Ford's Les Powell and Harry Firth created an upgraded version of the Cortina GT, called the GT 500.

Because the GT 500 was to be manufactured in Australia, the Series Production rules required that only 100 examples needed to be built. Cortinas dominated the race, though they suffered a number of failures and accidents. After 500 miles, GT 500s filled the top two places, with Barry Seton and Midge Bosworth declared the winners.

The rules were changed for 1966, increasing the number of homologation specials required to be produced from 100 to 250 cars. This immediately outlawed both the Cortina GT 500, and the 1966 race was dominated by a swarm of Mini Coopers which had little opposition.

For 1967, however, Ford returned to the Mountain, not with a hotted-up British compact, but with its own locally manufactured Falcon model. The Falcon had been in production in Australia since 1960, but it wasn't until the release of the new 'Mustang bred' XR in August 1966 that Ford start producing a true high-performance version designed to win Bathurst.

HO

351-GT

As well as two different-sized inline-6 engines, the new XR Falcon was made available for the first time with a V8, and it was this potent addition of a bent-eight that formed the basis for Ford Australia's first home-grown muscle car, the Falcon GT. In standard form, the 289 cu inch Windsor unit produced 200 horsepower, but with increased compression and hot camshaft, this increased to 225 horsepower for the GT. In addition it had a 4-speed manual gearbox and 2.93:1 final drive ratio. Disc brakes were fitted to the front with 11-inch rotors, and after having metallic gold paint applied (all XR Falcon GTs were painted gold except two special-order white models) and black stripes added, the exciting new Falcon GT was a sensation.

In total, just under 600 XR Falcon GTs were built, and they dominated Bathurst in 1967, taking six of the top ten positions in qualifying, and the top two places in the race, with Harry Firth and Fred Gibson driving the winning car.

Not to be outshone, Holden traded blows in 1968 with the arrival of its striking new Monaro GTS327 two-door coupé, while Ford upped the stakes with its XT Falcon GT, now sporting a larger 302 cu inch V8. However, Holden won the battle this time around, taking the top three spots in qualifying and in the race – and thus a horsepower war ensued.

For 1969, Ford released its brutish new XW Falcon GTHO, a homologated variant of its latest GT. A mighty 351 cu inch V8, a plethora of striping and new 'Superoo' graphics featuring a motorised Kangaroo with wheels and exhaust pipes were all standard features on the new GT, as were racing-style bonnet pins, plus a driver's-side bonnet scoop, aimed at ducting cool air to the brake master cylinder. However, impressive as the GT was as a road car, the newly introduced GTHO was built specifically for racing, and to win Bathurst.

As additions to the GT, the GTHO featured a front chin spoiler, stiffer front springs, stiffer front anti-roll bar, plus a rear anti-roll bar. For powering up Bathurst's Mountain Straight and back down Conrod Straight, it utilised a 3:1 final drive ratio, rather than the GT's 3.25:1.

As the Bathurst race was still broken into classes based on retail price, the GTHO was specifically priced to fit Class D, the same class in which rival Holden raced. Holden, not to be outdone, also opted to go bigger with its new HT Monaro GTS350, with its larger 350 cu inch engine.

Ford probably should have won Bathurst in 1969. A Falcon GTHO qualified on pole, and GTHOs took six of the top ten qualifying positions. But a multi-car shunt in the first lap, plus ongoing tyre failures stymied the Ford team's efforts, and while its brightest new star Allan Moffat was in contention, he was called into the pits late in the race to change tyres, as a preventative measure, which ultimately proved unnecessary as he'd already been nursing his tyres. The stop dropped him down the order and Colin Bond and Tony Roberts, driving for the newly established Holden Dealer Team, took victory in their Monaro.

For 1970, Holden chose to go small and nimble, and chose to develop its new Torana, which originated as a slightly stretched and restyled Vauxhall Viva. Now sporting an inline-6 cylinder engine from the full-size Holden range, the new Torana lacked the outright power of the V8 Monaro, but was more nimble, more economical, and with better handling and stopping power.

Meanwhile, Ford continued its development of the Falcon GT, with the release of the GTHO Phase II in August 1970. Still based on the XW model, and outwardly appearing quite similar to the Phase I, beneath the skin there were some notable changes. The most obvious of these was replacement of the Windsor-based motor with the new Cleveland. Although still rated at 351 cu inches, the Cleveland could rev 1000 rpm higher than the Windsor. The Phase II was fitted with a revised close-ratio 4-speed gearbox, Detroit Locker 9-inch diff, and 31 spline rear axles. Final drive ratio was 3.5:1. To improve braking, the front dust shields were removed, while wider finned drums were added to the rear.

Having lost Bathurst in both 1968 and 1969 to rivals Holden, Ford finally returned to victory in 1970, with the Phase II completely bullying the new Torana. Falcons took the top three places in qualifying, and

seven of the top ten positions on the grid. Come the race, Canadian-born driver Allan Moffat won the first of his four Bathurst victories, and Bruce McPhee in another Falcon finished second. It was an emphatic victory.

However, Ford was not done yet. Knowing full well that Holden would come out swinging in 1971, and with Chrysler Australia now entering the battle with its stylish new Valiant E38 Charger coupé, Ford needed to go better again. Its efforts would culminate in what is considered the greatest Australian muscle car of all time: the Falcon GTHO Phase III.

The Phase III was based on the GT version of the face-lifted XY model. It continued with the successful Cleveland 351 cu inch V8, now producing over 370 horsepower. Whereas the Phase I and II both featured a driver's-side duct fitted to the bonnet that directed cool air to the brake master cylinder, the Phase III incorporated a central duct attached to the top of the engine, and poking up through a hole in the bonnet. Nicknamed the 'shaker', the scoop came from Ford's US division, originally fitted to the Mustang, and helped force cool air into the 351's thirsty 780cfm Holley carburettor.

Like its older siblings, the Phase III wore a front spoiler, which was now teamed with a boot-mounted adjustable rear spoiler, also from the Mustang. Mechanicals were beefed up, a larger radiator was fitted, while rear brake drums were further widened. Larger front and rear anti-sway bars were fitted. At the time of its release in September 1971, the XY Falcon GTHO Phase III was believed to be the world's fastest mass-produced sedan.

Come Bathurst, Ford's ultimate supercar trumped the opposition. The first seven positions in qualifying were taken by GTHO Phase IIIs, and two E38 Chargers were all that stopped Ford from filling all first ten grid places. Come the race, Moffat again powered home in first place, followed by the GTHOs of Phil Barnes/Bob Skelton, and of David McKay. Colin Bond's XU-1 Torana in fourth place was the only non-GTHO in the top six. It was a total whitewash, and as history would show, the mighty GTHO Phase III would be the last all-out effort by Ford to produce the ultimate production car with the intention of winning Bathurst.

In mid-1972 Ford's construction of its latest Bathurst weapon – the new XA-based Falcon GTHO Phase IV – was well underway when a media outburst over manufacturers building high-performance supercars that anyone could purchase caused something of a scandal. As a result of this, Ford, Holden and Chrysler all quickly cancelled plans for their upcoming models. Rumours were rife as to what Holden and Chrysler were building, but there was talk of V8-powered Toranas and Chargers. However, these ultimately came to nought.

Ford pressed the XY GTHO Phase III back into service for the 1972 Bathurst race, albeit after hurriedly making available 15-inch Bathurst Globe wheels to replace the 14-inch 12-slot wheels the model was originally released with. These Globe mags were sent to GTHO owners free of charge by Ford so that it could prove enough sets were available on production vehicles.

Meanwhile, Holden upgraded the Torana XU-1 with a larger motor and made other improvements, while Chrysler rid the Charger of its main Achilles heel from 1971, its 3-speed gearbox. The new Charger E49 for 1972 was fitted with a 4-speed.

Once again, GTHOs dominated qualifying at Bathurst, taking the top four positions. Come the race, however, the heavens opened, and the smaller, nimble Toranas were in their element. The race was ultimately won by the Holden Dealer Team's fresh-faced youngster Peter Brock, who took the first of what would ultimately be nine Bathurst victories. The first GTHO home was the purple John French machine in second place.

Following the 1972 'supercar scare', and the bad press that came with it, manufacturers were understandably a little gun-shy about building cars like the GTHOs, which were essentially race cars that joe-public could own. Recognising this, and seeing the very real potential for manufacturers to turn their back on the sport, the Confederation of Australian Motor Sport (CAMS) moved quickly to bring about wholesale changes to touring car regulations, beginning in 1973. In one fell swoop, CAMS canned both Improved Production and Series Production regulations. In their place came Group C and Sports Sedans.

Group C incorporated many of the key aspects of Improved Production and Series Production. Much as for Series Production, the cars needed to be of a type either manufactured in Australia or sold in significant numbers in Australia. Manufacturers still needed to produce road cars from which key features could be carried across to the benefit of the race cars. But the rules allowed many more freedoms than Series Production. Items such as wheels, tyres, specific brake parts, exhaust systems, carburettors, shocks and springs, fuel tank, driver's seat, steering wheel, and various other parts could be replaced with purpose-built racing components.

This all meant it was no longer necessary to create such purpose-built cars as the GTHO, whose road car parts would be put to test on the race track, and therefore needed to be heavy duty and/or high performance by nature. Whereas in Series Production the road cars and race cars had been one and the same, in Group C the two were completely separate. And while the race car needed to evolve from the road car, and race fans could recognise the vehicles as being the same make and model they themselves drove on the street, the two were quite obviously very different, which removed any chances of there being a repeat of the 1972 supercar scare.

This also meant that the Falcon XY GTHO Phase III was the last in the line from a fascinating chapter in Australian motor racing and car manufacturing. By all accounts, this car became an instant legend and an instant classic. Its mystique and its significance held it aloft, and it has always been a hugely desirable vehicle. Since its release in 1971, the GTHO Phase III has led the way among classic Australian cars in both collectability and value. It has never been possible to pick up a GTHO Phase III for a bargain price. And that continues to this day. The past decade has seen prices paid for classic Australian muscle cars soar, with the Phase III in the forefront of this trend – some vehicles changing hands in Australia for in excess of $1 million.

In New Zealand, only a handful of GTHOs were imported brand new. It's thought the total number was about seven cars. Most of these cars were imported for the same reason they were built in the first place:

to go motor racing. The new Castrol GTX production car racing series allowed the Australian supercars such as the GTHO, Torana XU-1, Charger E38 and later the E49 to compete. And many of these makes and models did just that. Jim Richards won the first Castrol GTX championship held during the 1971/72 season driving a GTHO Phase III, while others such as Robbie Francevic, Richard Brocklehurst, Bruce Anderson, and Patrick Smith also piloted GTHOs.

In fact, it's actually quite rare to find a New Zealand-new GTHO Phase III that *wasn't* circuit-raced when new. But the Vermillion Fire with black trim example featured here is just that. Long-time owner Rob bought the car in the mid-1980s. 'I bought it approximately 32 years ago', explains Rob. 'It was only slightly more expensive than our first house. It was sold new in August 1971 to a buyer in Taupo. It was drag raced when it was one year old, it still has a New Zealand record. It's the only Phase III to come new to New Zealand that wasn't circuit-raced.'

Rob has exercised the GTHO plenty over the years, doing sprints and the like, but such is the model's significance, even when he first purchased it in the 1980s, that he always knew what an important car this was. And as such, it has always lived a charmed life, much more so these days.

Rob also owns a GTHO Phase I, although this car has no New Zealand history. It's a car he imported from Tasmania. 'The man I purchased it off had two Phase Is and a Phase III. I would love to get a Phase II, but would have to win Lotto to get one these days.'

Of the original New Zealand new Phase IIIs, at least one has since been sold back to Australia. Those that remain are highly prized, and on the odd occasion they're displayed at car shows, quite rightly generate enormous interest.

2
Castrol GTX
Castrol
Castrol GTX
GRACE BROS
2
Castrol
TAA
2
GRACE BROS
TOYOTA

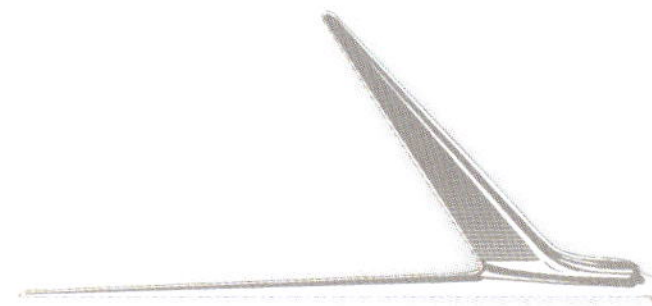

PETE GEOGHEGAN'S SUPER FALCON

The last person ever to win an Australian Supercars Championship race in a Ford Falcon was Kiwi phenomenon, Scott McLaughlin. McLaughlin's victory came in Race 1 of the final round of the 2018 Championship.

The Supercars moniker is essentially a branding name. But Supercars itself is simply the latest evolution of the Australian Touring Car Championship, which began in 1960. The reason McLaughlin's victory was the last ever for the Falcon, is that 2018 was the last year the model was eligible to compete in the series. Indeed, Ford had stopped producing the Falcon two years earlier.

While McLaughlin will forever be etched in history as having scored the last Falcon victory, the very first was achieved by another racing legend, Ian 'Pete' Geoghegan.

Geoghegan had a relatively short motor racing career, stepping away from full-time racing when he was just 32. Nevertheless, he'd already won the Australian Touring Car Championship five times. To this day, only Dick Johnson, Mark Skaife, and Jamie Whincup have won five or more championships, and all three of them raced well beyond the age of 32.

Geoghegan came from a family of racers. His brother Leo was also extremely talented, although he preferred single seaters and sports cars over touring cars for much of his career. The Geoghegan family had a well-known car dealership in Sydney called Geoghegan's Sporty Cars, which sold mostly high-end vehicles, specialising in Lotus.

From 1960 through to 1968, the Australian Touring Car

A rare sighting of the Super Falcon built for Allan Moffat. It was hauled to several Australian Touring Car Championship events in 1971, but invariably, Moffat preferred to race his famous Kar-Kraft built 1969 Boss 302 Mustang.

Championship (ATCC) was contested as a single race event. The venue changed from year to year, and while teams and drivers kept themselves busy racing right throughout the season, it wasn't until 1969 the championship was conducted over several races.

Geoghegan rose to prominence in the late 1950s driving a Holden FX, but it wasn't until he switched to a MkI Jaguar that he became a genuine outright contender in the ATCC. In 1961 he placed second to Bill Pitt's similar machine at the Lowood circuit, in Queensland.

In 1964, driving a Cortina GT, Geoghegan won his first Australian Touring Car Championship, which was held at Lakeside. In a battle that lasted the entire duration of the race, he fought such legends as Bob Jane (MKII Jaguar), Norm Beechey and Brian Muir (driving EH Holdens), and Jim McKeown (Lotus Cortina), to finally take the victory by just 1 second over Beechey.

1965 saw the arrival of the American muscle cars as true outright contenders, and Geoghegan headed to the US to purchased himself a Mustang, as did Beechey and Jane. Beechey was the first person in Australia to race the new Mustang model, and he ultimately won the 1965 ATCC, while Geoghegan again ran a Cortina, while his own Mustang was being race prepared.

However, as soon as the Geoghegan Mustang hit the track it was a winner. Geoghegan won an endless number of races throughout the remainder of 1965, and right through until 1967. Along the way he also scored his second ATCC, in 1966, which was held at Bathurst. He battled for much of the race with Beechey's latest American import, a 327 cu inch Chevy Nova, but ultimately wore down his rival as the big Chev ran out of brakes.

The tables were turned for 1967, as it was Geoghegan who now had the brand-new untested car, when the ATCC took place at Lakeside in Queensland. Beechey had had a year of development on the Nova, and was comfortably in control of the race when a tyre blew and he was sent careering into the Armco. Geoghegan, who was running second at the time, cruised to victory, and ATCC number three.

Geoghegan's new Mustang GTA was continuously developed into the fastest racing sedan in Australia by the end of 1967, and although Beechey had imported a new 350 cu inch Chevy Camaro, and Bob Jane had a new Mustang, Geoghegan's fourth ATCC in 1968 appeared relatively easy, as he powered away from his rivals in the early laps, before they all broke.

1969 was the first year the ATCC was contested as a multi-round championship. There were to be five races, and Geoghegan had all but locked up the title after the first four. Alan Hamilton, driving a Porsche, came close to stealing the crown from him, but needed to win the final race to do so. Geoghegan was disqualified for the final after his car failed to start just prior to the race, and he didn't manage to get to the grid in the required one minute before the flag dropped. Ultimately, however, Hamilton came up just short, and could only muster second in the Porsche, so Geoghegan locked in his fifth championship – three of which were achieved in the same car.

The Geoghegan family had close manufacturer ties, particularly with Ford, and brothers Ian and Leo often shared a factory-supported car at Bathurst each year. In the early to mid-1960s, it was Cortinas, before the new Falcon GT arrived in 1967. And it was from these manufacturer connections that the replacement for Ian's now elderly Mustang GTA would emerge in late 1970.

In 1970, Norm Beechey won the ATCC driving his epic new HT Holden Monaro GTS350. Holden supported Beechey's racing programme, including that in 1969 with an HK Monaro GTS327. The ATCC was contested to Improved Production regulations, which allowed a range of mechanical freedoms over the standard production vehicle. Beechey developed the HK Monaro throughout 1969, and claimed two race victories, at Surfers Paradise and at Symmons Plains. Indeed, the Surfers Paradise win was the very first time a Holden had won an ATCC race.

For 1970, Beechey went all-out with his new HT Monaro, and scored three victories from the seven-round series, to lock up the title with one race remaining. While his HK Monaro was the first to score an ATCC race win in 1969, the HT was the first to win the championship.

The ATCC and Improved Production racing as a whole enjoyed massive support, but while Holden was basking in the glory of its 1970 victory, Ford Australia had little presence. The annual Bathurst endurance event was contested to Series Production regulations, which essentially required the cars to remain completely stock-standard. And while Ford's Falcon had been a contender for victory every year since the debut of the XR GT in 1967, Bathurst was just one race. Ford needed to keep the Falcon at the forefront of Australian competition throughout the year. It needed a contender in the ATCC. And even though Mustangs had won the championship every year from 1965 through to 1969, Holden had one-upped its rival by winning the championship with a home-grown model. Mustang race victories didn't necessarily help to sell Ford Falcons, whereas Beechey's successes in his Monaro helped to sell Holdens. To that end, Ford Australia launched an aggressive programme to build a pair of heavily modified XW Falcon GTHO Improved Production cars for both Geoghegan and fellow Ford factory driver Allan Moffat.

Although Pete Geoghegan's Super Falcon was painted in the yellow and green colours of Grace Brothers when it appeared in New Zealand in late 1972, for much of its Australia career, it was painted in the white and green of his long-time sponsor, Castrol.

The plan was to contest the 1971 ATCC with the Falcons, and American Al Turner, Ford Australia's racing boss, rubber-stamped a construction budget of A$160,000 (well over $1 million in today's money) for the two 'Super' Falcons. Given that a brand new GTHO Falcon cost less than A$4500, this was a sizeable amount. But it still wasn't enough. In the end, the price of construction of the Super Falcons more than doubled.

The brief for the two Super Falcons was substantial weight reduction and epic power. Most body parts were acid dipped. This involved marinating the parts in a large tank of acid, which eats away at the metal and reduces its weight. The result was body panels that were paper thin, and required delicate handling when opening and closing the doors, bonnet, and boot. The Falcons would wear

15 x 10-inch magnesium Minilite wheels, the maximum allowed under Improved Production regulations. They'd be wrapped in fat Goodyear tyres, requiring the wheel arches to be flared 3-inches to house them.

The bumpers were stamped in thin-gauge metal, and much of the front inner structure was removed. Headlights, bumpers, grill, and other bright work were attached to fabricated brackets. Magnesium was lavished upon the Falcons to save more weight, right down to the wheel nuts.

All-up, the two Super Falcons weighed just 1315 kilograms, which was nearly 200 kilograms lighter than an XW Falcon GT street car. This was impressive given the Improved Production regulations required almost everything from the street car be carried across to the race car, including the carpet! In addition, the race cars were also fitted with roll cages. In the case of the Super Falcons, these were made from aluminium.

The Super Falcon engines were works of art. They used the Cleveland V8, with modified Falconer and Dunn cylinder heads sporting alloy roller rockers and titanium valves. Internals included a steel billet crankshaft, special pistons designed for 130-octane fuel, and conrods machined from Repco Formula 1 castings. Dry-sump systems were also included, making these some of the very first racing sedans anywhere in the world to be so fitted. Atop the engines sat a mechanical fuel-injection system. Spent exhaust gases exited via four-into-two headers, connected by a balance pipe. The exhaust pipes measured 4 inches in diameter, and of course, racing cars weren't fitted with mufflers in 1971.

Initial power figures showed 570 horsepower, but this increased to around 620 by 1972, an astonishing figure for the time.

Much interest surrounded the two new Super Falcons, to which the Australian motoring press had cottoned on early in the programme, long before the cars were unveiled to the public. Allan Moffat's example appeared for the final round of the 1970 Australian Touring Car Championship, where he completed four smoky laps before the engine blew itself apart.

By the time Moffat's Super Falcon made its next competition appearance in Round 2 of the 1971 ATCC at Calder Park, it'd been upgraded with the latest XY front. Both cars missed Round 1 due to ongoing delays and setbacks in the development programme.

In 1971, Moffat was racing a state-of-the-art 1969 Mustang built for the Sports Car Club of America (SCCA) Trans-Am series by Kar-Kraft. Kar-Kraft worked with Ford Motor Company in several high-profile racing programmes, including the GT40 at Le Mans. After two Le Mans victories in 1966 and 1967, Ford concentrated on the Trans-Am series. Having been a faithful warrior for the company, Moffat was gifted one of the 1969 Kar-Kraft factory Mustangs to take to Australia. Despite its little 302 engine being a good 100 horsepower shy of his Super Falcon, the brilliant Mustang was a far superior car to the brutal Aussie sedan. Indeed, as the 1971 season progressed, Moffat found himself in contention to win the championship, engaging in a slugfest with the mighty 7-litre ZL/1 big-block Camaro of Bob Jane.

The two Super Falcons appeared together for the first time in Round 3 of the ATCC at Sandown. They were painted to match the pair's Mustangs: Moffat's in red with black stripes (for his sponsor Coca-Cola) and Geoghegan's white with green stripes (for his sponsor Castrol Oil). Both cars practised, and Geoghegan ran his in a preliminary race, but both drivers switched to their dependable Mustangs for the all-important ATCC race.

Geoghegan took only his Mustang to Surfers Paradise for Round 5, while Moffat took both his Mustang and the Super Falcon. By now, it was clear the Mustang was his best chance at winning his first championship, and for this reason, John French drove the Falcon. While Moffat won from Bob Jane (Camaro) and Geoghegan, French battled the white Mustang throughout for the final podium position.

Queenslander French was drafted in to drive Moffat's Super Falcon again in Round 6 at Lakeside, and finished fifth behind Moffat, Geoghegan, Jane, and Norm Beechey's Monaro.

With both Moffat and Geoghegan in the hunt to win the championship, neither bothered even practising their Super Falcons in

Geoghegan sits on the front row of the Bay Park grid in late 1972, alongside Frank Gardner's very quick 1967 Camaro, which he'd brought from England. The gold Camaro on row 2 is that of Kiwi racer Red Dawson.

A packed crowd was in attendance to see Geoghegan take a popular race victory at Bay Park, a rare sight in the troublesome Super Falcon.

the final round at Oran Park. Moffat would have won the ATCC had his gearbox not stuck in gear while he was leading the race. He stopped to resolve the issue, dropping back to third behind Jane and Geoghegan, and began a mighty pursuit that took him right to Jane's rear bumper by the final lap, but the Camaro driver stood firm, and claimed his third championship.

Moffat's Mustang was every bit the equal of Jane's Camaro, and he opted to abandon his Super Falcon for 1972. Geoghegan, however, didn't have that luxury. His Mustang first began racing in 1967, and despite countless upgrades, was past its use-by date. He needed to make his Super Falcon a race winner.

To that end, Geoghegan's Super Falcon was sent to specialist sports car builder Bowin Racing Cars to be completely rebuilt and redeveloped. Most notable changes included a new roll cage structure to make the car significantly more rigid than the baggy mess it was in 1971. In addition, even more power was extracted from the 351 Cleveland. When it emerged in early 1972, Geoghegan's Super Falcon bristled with a monstrous 620 horsepower.

Although the upgrades to the Super Falcon were significant, the competition hadn't stood still either. New regulations for 1972 forced Bob Jane to fit a 350 cu. inch small block to his Camaro, and again he and Moffat went at it. And once again, the Chev driver emerged as the champion. Indeed, there were only three race winners in 1972: Jane, Moffat, and Geoghegan. Jane won four, Moffat three, and Geoghegan one. The Super Falcon still wasn't a match for the smaller, more purposeful American pony cars, but it was now significantly better than Norm Beechey's old Monaro.

Geoghegan retired from Round 1 of the 1972 ATCC with an overheating engine, while he qualified and finished third in Round 2, despite pitting to check an oil leak. Everything came together, finally, in Round 3 at Bathurst, where the wide open spaces and long straights allowed the Super Falcon to be fully tapped. The early laps saw Geoghegan battle Jane and Moffat for the lead, before the Camaro broke. That just left the two Fords. Geoghegan was in front, and scorched away up Mountain Straight and back down Conrod Straight, while Moffat came hurtling back up again through the bendy stuff, and sometimes even pulled ahead briefly. Then the Falcon began dumping oil, leaving a smoke trail in its wake, much of which ended up smeared across Moffat's windscreen. He unbuckled his racing harness and stuck his head out the side window so he could see, and kept the pressure on. In the end, however, it was Geoghegan who was victorious in the 13-lap epic. It was a battle of the greats, soured somewhat when Moffat lodged a protest, claiming his rival should have been black-flagged for the oil spray. In the end, Geoghegan kept his win. Nearly half a century later, this still ranks as one of the greatest races in ATCC history. And somewhat lost in all the excitement was that this was the very first ATCC victory for the Ford Falcon.

The remainder of the 1972 season saw Geoghegan score a couple of second places, but, overall, the Super Falcon hadn't lived up to expectations. However, the car proved massively popular among spectators, and Geoghegan's incredible driving skills saw the big machine crossed up and sideways much of the time, laying down a pair of elevens on the track as it did so. It was the most powerful car on the grid, but not even close to being the most refined. Despite the colossal effort, Bathurst was to be the Super Falcon's one and only ATCC victory. But what a victory it was.

For 1973, the Confederation of Australian Motor Sport changed the regulations under which the ATCC would be contested. The popular Improved Production cars would be outlawed, and thus the Super Falcon could no longer challenge for the series it was created for. However, although he'd sell the car in 1973, Geoghegan was invited to bring the Super Falcon to New Zealand, to race at the Bay Park Christmas meeting, and in the saloon car support races at the New Zealand International Grand Prix at Pukekohe one week later.

In addition to Geoghegan was Moffat and his Mustang, along with international Australian racing driver Frank Gardner, out from the UK in a Camaro. American Trans-Am independent Joe Chamberlain would also make the trip with another Camaro. Up against this impressive international arsenal were the best New Zealand big banger racing sedans.

At Bay Park, Moffat took pole, followed closely by Gardner, while the spectacular Geoghegan clocked the third-fastest qualifying time. The tight confines of the Bay Park circuit didn't suit the hugely powerful Super Falcon, and Geoghegan used every inch of tarmac and much of the gravel and dirt that lined it just to extract a competitive time. He fed in massive amounts of wheel spin exiting each corner and clocked the spectacular Falcon in a near-constant drift, spitting dirt and stones at amazed onlookers as the outside rear tyre exceeded track limits. It was fast, but more than this, it was enormously entertaining.

In the first race, Geoghegan used the Falcon's epic power to muscle past Gardner from the start, and held him there for much of the race while Moffat sped off into the distance. Gardner eventually got through, but Geoghegan put up a fight. Gardner dropped all the way back to fourth in Race 2, with Moffat again in front, followed by the superb Geoghegan, and Kiwi driver Red Dawson in a Camaro. But Gardner worked his way back past Dawson, and while it took him several attempts, he eventually also displaced Geoghegan. However, having caught Moffat, Gardner got desperate, and went careering off backwards into the wooden catch fence.

The final race saw Gardner a non-starter, and once again Moffat was off and gone while Geoghegan treated the huge crowd to a dazzling display of car control as he wrestled the unwieldy, overpowered Super Falcon into submission. It appeared Moffat would make a clean sweep, but suddenly the Mustang slowed, as a split radiator hose saw the temperatures soar. Moffat pitted, Geoghegan swept through into the lead – and as he did so, the crowd erupted with joy. This would be an enormously popular victory. Winning while also entertaining was truly appreciated.

One week after Bay Park, the teams moved on to Pukekohe for the Grand Prix meeting, and disappointingly, the saloon cars got just one race. Once again Moffat qualified fastest, followed by Gardner and Geoghegan. Pukekohe had had two new chicanes installed following a fatality at the Grand Prix the previous year, and most drivers were treating them with respect. But not Geoghegan. He threaded the big Falcon through them with confidence and bravado, flicking one way, then the other, before mashing the accelerator and hurtling away.

Come the race, Moffat took off, just as he had at Bay Park a week earlier. But Gardner made sure he stayed close to the Mustang, while Geoghegan shadowed the Camaro. But in applying pressure, the Super Falcon driver clipped Gardner through one of the chicanes, and before he could react, he was slammed straight into the immovable tyre and concrete structure. The front of the Super Falcon was badly bent, and Geoghegan headed for the pits for a little makeshift panel beating before re-joining the race.

By the time he resumed, Geoghegan was well out of contention, but he banged in a series of hot laps, and drove the brutal Super Falcon with its banana-shaped bumper as if he was contesting the race victory. For the appreciative Kiwi crowds, watching the spectacle of this amazing Ford Australia factory-backed muscle car and its spectacularly gifted driver was something to savour.

Geoghegan did return to New Zealand a couple of years later, though not in the Falcon, but instead in a Porsche 911. The Falcon was sold in 1973, and raced in the Australian Sports Sedan category with its new owner Kingsley Hibbard. However, it was never a front runner again, and was largely forgotten until being tracked down and purchased by Queensland enthusiast and collector David Bowden, who had it restored back to its all-important 1972 Improved Production look. Its New Zealand appearance was brief, but oh-so spectacular.

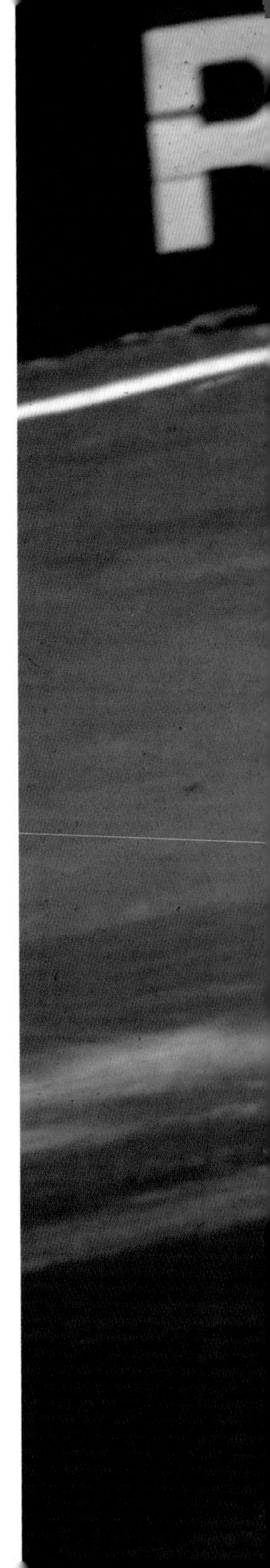

Huge focus went into making the Super Falcons light, and that included stamping bumpers made from thin-gauge metal. When Geoghegan went clattering into one of the new Pukekohe chicanes in his second New Zealand outing, the bumper didn't stand a chance. Geoghegan spent several laps in the pits having the bodywork pulled from the tyre, then went out and punched in some spectacular laps for the big Kiwi crowd.

Sean McCaughan races this beautiful Pete Geoghegan Super Falcon tribute in the Historic Muscle & Saloon Cars group.

Left: A great-looking Kiwi tribute to an Australian race car – Allan Moffat's famous 1971 Bathurst-500-winning XY GTHO Phase III.
Below: Mike McCaughan's XY GTHO Phase III replica Historic Muscle Car at Hampton Downs.

Above: Ross Tebbs' XY GTHO Phase III.
Left: Garden art at Rob's place.

US versus Australia Ford battle on the Waiouru military airstrip. 1969 Boss 302 Mustang in the near lane against XY GT Falcon in the far lane.

GZ 2668

JAMES BERRYMAN

James Berryman has, quite possibly, owned more Falcons than has anyone else in New Zealand, barring a Ford dealer. In all, his tally of Falcons owned exceeds three hundred (300!) cars! These include everything from GTs, GSs, Fairmonts, and utes, right through to base 6-cylinder models. Indeed, the number of GTs that have passed through his hands stands at 17.

Of course, had he kept them all he'd have a multi-million-dollar car collection now. But his was never a long-term plan. One car was sold to fund the next. They came and went, and some were wrecked along the way.

James is the son of legendary speedway racer 'Jumbo' Berryman. His stories are typical of so many fast young Kiwi blokes during the 1970s, '80s, and '90s, who owned and enjoyed – and thrashed – what are now highly desirable Aussie muscle cars. Of course, they weren't that back then: at the time, they were simply cheap and fun transport, good for doing burnouts and pub crawls.

Fortunately, James took a lot of photos of his cars, and recalls a lot of stories, which he shares on his excellent Facebook page, 'You Had To Be There'. Here are a few samples.

1: Palmerston North about the year 2000

I spotted an advertisement for an XA Fairmont coupé down in Levin. I got my mate Pom to give the chap a call and he was told that his partner was due to have a baby and she thought it was time to get rid of the old coupé and buy a nice practical family sedan ...

At this time I had a couple of VL Commodores for sale so we hatched a plan to go look at the coupé in the VL Commodore and try to offer a deal. We met with the chap in Levin and you could tell he already had his life sorted out and nicely organised by his lovely partner. I don't think he actually wanted to part with his Fairmont coupé, but not long after this we handed him the VL keys and $2100 cash, and we headed for home.

This Fairmont coupé was the chap's dream car and probably the tidiest original condition coupé I've had. With 80,000 miles and as picked up in the pictures (I had not even washed it), it was a beautiful thing.

I often think about that chap and hope his lovely partner has let him replace the void we saw in his life as we drove away. I do doubt he would be able to buy a nice Fairmont coupé with the VL Commodore and a couple of grand cash nowadays.

I also know how kids turning 19 or 20 years old are nowadays and how much they appreciate what sacrifices were made for them.

2: Palmerston North around 1999 or 2000-ish

I get a lot of people commenting, asking the same old thing: 'Bet you wish you still had that?', or, 'Do you regret selling it?' etc. etc. ... Well, not really, because money was always tight and I was constantly selling one to buy the next ...

In the good old days of living with my Falcon obsession, I have owned over 300 1967 through to 1987 Falcons, Fairmonts and Fairlanes, so I never really got too attached to any one Falcon.

Here is a fairly good way to explain how things happened. I got these two GTs about the same time for a bargain price of $5000 each. I was reading the local *Manawatu Standard* newspaper one day and saw a little advertisement in the weekly free column where people sell stuff like old coffee tables, Playstations and crap ...

All the little advertisement said was, 'Red Ford V8. As is. $5500 ono.'

It was listed under a local rural phone number so I thought I'd call and see what model it was. The lady said she didn't know but it was very noisy and old. She said her boarder was working over in Australia and had asked her to see if she could advertise it just in case someone might be interested.

I drove straight out to the farm address and there it was. A factory XA GT. It had the factory wind-back sunroof and 4-speed manual.

Around the same time my old mate Ian Humphrey had recently opened his new car yard a few doors down from my shop that I had also recently opened. We got on well and I was buying a lot of older model Falcons from him that were traded in at his yard.

Ian called in and said that he was trading an older model V8 Falcon. He also said that he was trading it in for $5000 and would I be interested? I was, but I was stocked up on Falcons at the moment. He said, 'No worries, just take it and sort it out when you can'.

I couldn't believe it when he called to ask me if I could pop down to the yard and pick up this old Falcon as he needed the room. It was a really nice original XB GT; factory wind-back sunroof, manual, power steering, and like a new car to drive. It was a little beauty ...

So, there is a good example of why the cars came and went like hot dinners. If I kept every one that I liked, I would probably only have my first car and that wouldn't be much to talk about. There was always another one coming along.

3: This one was my first Fairmont GS

In 1988, I picked up this tidy old GS Fairmont body from down South with my mate Orifice. We flat-towed with a rope; the old thing with Orifice's old XY 6 cyl at full speed through the day and night, back North.

This was my first full car building experience and I was a little green on the knowledge regarding engines and gearing, etc.

The first engine I bought was a rebuilt 302 Cleveland from a mate.

RF4715

GT351

MJ 3108

Acting against my father's advice that a 302 Cleveland fitted with open chamber 351 heads and a cam grind suited to a Don Garlits Top Fueller ... we got it up and running. It sounded like a 6-second drag car but had about as much torque and power as a Daihatsu Charade.

The first and only time I drove it with this combination was on a trip up to Blenheim with Innocent Ashley and Orifice one weekend. Innocent Ashley was the King of Demolition Derbys in the good old days and between Woodford Glen and Nelson and any other track, he won at least six derbys in a row, all the time piloting the trusty Phase III Vanguards.

The Vanguard was one that he had just taken up to Nelson and he'd won the first place. The prize money in those days was well worth the effort.

I was working for Ashley at this time and we heard about a derby the week after Nelson, up in Blenheim, so it was decided that a road trip, in the recently finished GS, would be a plan.

Ash reckoned that the Vanguard was healthy enough after the Nelson win to take out another derby. He offered to pay for the gas and accommodation which was a bad mistake from the start.

That 302 Cleveland loaded up with the Vanguard on the trailer was the most anaemic, uneconomical piece of crap ever. When we finally arrived at Blenheim after our third fill of gas, and first gear around the hills and over the pass for miles, we all agreed that it was not going to be a profitable trip even if he did win the derby. We booked in at Bings Motel, went to the Speedway and the old Vanguard was good until the very end. It died with Innocent Ashley finishing second.

After a big night on the turps we were on the road to travel through the night back to Christchurch. We were dreading the long slow journey ahead in the Fairmont with the beaten-up Vanguard on board. But things took a strange turn when we finally got to the top of the hills just out of Blenheim. We could not work out why but, for some reason, the old 302 felt like it had double the amount of power suddenly, and we were going along through the hills at a great pace. We made it back to Christchurch in about half the time it took to get up to Blenheim and we used just one tank of gas. The difference was unbelievable.

Anyway, we arrived back late in the night and as we pulled up Innocent Ashley was greeted by a very outraged Aunty Sandy who had been waiting up for us. She was going right-off saying that she had had a call from the Blenheim Club asking if we had lost our Vanguard off the side of the highest cliff just outside Blenheim. Unbeknown to us, the bloody thing had come off the trailer and launched off into the night off the cliff.

Ash, Orifice and myself were speechless and just couldn't believe how it could have happened ... but it did explain the sudden increase in power and amazing economy on the way home!

Not long after this terrible run of bad luck, I got my mate Morph (R.I.P.) to build a healthy 351c with some good gear and 9-inch diff from Cwazy Cwris Leigh. Once it was going I lost my licence, and also moved North to Auckland.

I had Jumbo drive it up to Auckland through the winter with no heater, wipers or demister. He turned up in Penrose with the worst attitude towards me after being roadblocked by the Police around the Blenheim area enquiring about cars going off cliffs and other lines of enquiry throughout the South. He had what looked like frostbite all over him from standing around in the snow and sleeping on the wharf after missing the ferry.

The last engine we fitted was a hot closed chamber 4V Falcon GT Motor, Top Loader and 9-inch limited slip diff in Auckland and sold to our old mate 'Numberplate' who carried on the tradition of chases and roadblocks all over Auckland.

4: Late 1980s

I used to hear about sightings and locations of good old cars all over the place and nine times out of ten, you were led to a dead end.

This one was a story about an original XY GT 'shaker' that was sitting in behind the building at the Papakura Dairy. I thought, 'I may as well go have a bit of an optic and snoop around the dairy, as you never know'. Well, you wouldn't believe it. The fat little Indian who had it parked up behind his dairy in Papakura agreed to sell me the old GT.

With a handshake and a pretty fair price of 10K, a jump-start and some fuel, the car was driven home.

It was a good old car. Soon after buying it, I drove it down to Christchurch and after a few nights out pubbing and partying, after my mate Dungeon attempted to put us in the Avon river backwards at speed, I decided to tart it up and sell it on.

After I had it painted at Pedo Panel and Paint and I had lost my licence for a change, I had Donut and Rugged Ronnie drive me up to Auckland to deal it off. It was a big trip North with the usual Cop stops, tickets and other fun events.

The poor old thing ended up broken. I traded it to a mad mate of ours up in Auckland and on its debut outing with the fresh pilot at the helm, we were involved in a good old cop chase around Bastion Point. After flipping it and sliding along on its roof with the cops in behind, our loopy crazy mate climbed out and with a big grin of satisfaction on his dial said to the cops, 'You boys order an H.O. to go?' Good times.

5: Palmerston North, around 1996

These were the best times when you had no online dealing and everything was easy.

One day I was looking through the old *Trade and Exchange* and saw this nice old XY GS Fairmont advertised in Christchurch. As it was running a nice big port 4V Cleveland, Top Loader and 9-inch diff, the price was fair at $7500. I rang the owner and the next day it was picked up, sight-unseen, and on its way on the truck to Palmerston North.

These were good days alright. It was a good old thing and went nice and was exactly as described by the seller. I fitted the 16 x 10-inch Globe mags all-round for a while before taking them off to sell to me old mate Pete Jacobs from 'No Hair Racing' for their XA Falcon coupé racer. These wheels really looked the part on the old girl.

Time came to move it on so, I advertised the Fairmont in the local *Trade and Exchange* magazine. A few days later I got a call from an automated phone system saying, 'This is a call from a prison inmate, press such and such to accept the call'.

I accepted the call and the chap said, 'Hey Bro, I want to buy that GS Fairmont. Can you bring it to Rangipo Prison on Saturday and if it's all good, I will buy it and get my Mrs to pay you and drop you back to Palmy after Visits.'

Well, straight away I recognised the voice from years gone by and knew who I was talking to so knew it would be all good. The Saturday came and my mate Geno and myself hit the road for Rangipo, just before Turangi. We arrived at the prison car park and, next minute, there was a tubby old prison officer waving us through the gates and straight through to the concrete area in the unit. My old mate walks over, has a laugh and smile, and the car was gone.

I suspect that it was a Blue-influenced unit as all the neighbours in the unit were happy as with the purchase of the blue Ford by my mad old mate. Good times and a little different to the visitation criteria nowadays.

6: Queenstown, late 1980s/early 1990s

I went on a road trip down to Queenstown with a mate, Mad Mike, in the old Falcon GT.

I pulled up at sunrise to pick Mike up for the trip; popped the trunk and said, 'Throw your bag in there mate'. He looked at me with disgust and said, 'What bag? I have everything we need here.'

About 50 kilometres before arriving in Queenstown, we had consumed all his luggage, which consisted of three 40-ounce bottles of rum and a couple of big bottles of Coke. The old GT was running sweet all the way.

At this time the old Stalker radar detector was giving out the occasional beep-beep with no sign of any Highway Patrol.

Then, in the distance we spotted the old black-and-white cop car parked on the side of the road. We slowed down to the right speed and as we came along we noticed the Officer was sitting on the hood of his patrol car and having a good old chat with the local Farmers.

So as you did back in the good old days, we gave them a wave, changed down a gear and as we went past, we wound it out to about 100 mph and into top gear and headed to Queenstown as fast as we could.

After a while we noticed the radar detector was beeping sporadically as we were going through the hills and corners. So, we slowed up in anticipation of seeing another patrol car, but nothing for miles. We then realised that it was actually the chatterbox Officer from way back in pursuit.

Then we came into the area where the 'Welcome To Queenstown' signage was and wouldn't you know it? Here was the local Queenstown Sheriff with a roadblock.

Man, he was all pumped up with anger and a negative attitude like we had never encountered before at a roadblock. (They were a regular thing.) He proceeded to tell us that we were not welcome in HIS town and that we could piss off out of the area. His reasoning was that we were wearing black jerseys, had long unkempt hair and the car was noisy and full of empty bottles. The sure signs of trouble, he said.

After a long chat and greasing him up, he decided we could stay for one night in HIS town at a paid accommodation venue and the car was to be parked until we left town. Plus, he assigned two full-time officers to follow us around on foot until our departure. He was one Power Crazy Sheriff, I must say.

Anyway, we had a great night walking bar to bar to club and were out of there at sunrise.

Crazy hospitality at the time.

7: Mid 1990s

Outside Boonge's pad on the side street in Greenlane. I got this nice old XB GT as transportation for the New Year's festivities one year. Bad

Head Barry had bought the XW Fairmont also to cart half the crew from Christchurch who had arrived to party.

These were great times when the old black-and-white M.O.T. cops were operating, as you could always get a good bit of banter and the odd chase out of them as they thought they were professional racecar pilots. Here they were just talking to us about the risks of driving with too many cans in the console.

This was a great holiday with mates providing entertainment everywhere, including The Fonz, here in the Bobby Policeman's helmet, Travelling Wilbury, Bungeye, Bad Head, Spoon, Travis Nine Years, Boonge, Magnet, Jim Carrey, Dave, Kerry, Dungeon and a few others.

Had a great New Year's all over the place and the old Falcon GT got a severe punishment which resulted in it having a bit of a bad name with the M.O.T. chaps, so I quickly sold it not long after this to our old mate Ming down in Hanmer who owned it for many, many years thereafter.

8: Late '80s, early '90s

My constant search for my next Falcon. At this stage of my life, I was consumed with road trips, racing, pissing up with my mates and just being a general Hood in my late teens.

In my search, I spotted this Factory XY GT Shaker at Turners Auction in Auckland.

This was the roughest Falcon GT I had ever seen, but it had all its original running gear, a few bits missing and a lot of spares in the trunk. It ticked all the boxes for what I needed. I got the final bid, picked it up and we got right into the preparation of what we required: the fastest party wagon in the land.

First off, we ripped the carpets out, front bumper and other unneeded weight. We bogged up the sunroof, stuck a half roll cage in it, added a lightened flywheel and twin-plate clutch, had a set of new headers made and a Bathurst-style side-exit exhaust with a 6-inch outlet. I took it to my good mate in Otago and he went over the engine which was pretty good already. We just fitted a full Sig Epson racing valve train, a new inlet, carb and ignition system.

Back to Auckland for some suspension, wheels and a few odd jobs and we were sorted.

I was over the moon to have the roughest, noisiest, toughest pub hack in the land.

It was indestructible and we used this thing hard. We had it clocked at 161 mph flat out on the open road and a 12.9 quarter mile with the road gearing.

It was always fully loaded with a team of Good Bastards on a mad mission. It was forever on many Gumball Rally-inspired piss trips from Auckland to Southland and back. The Donut eaters in the Capital in their XF and EA patrol vehicles loved us so much that they usually had a roadblock in place to welcome us to their area before we hit the Wellington to Picton Ferry Terminal.

These were good times back in the good old days and many a mate gathered on those travels.

PARTS

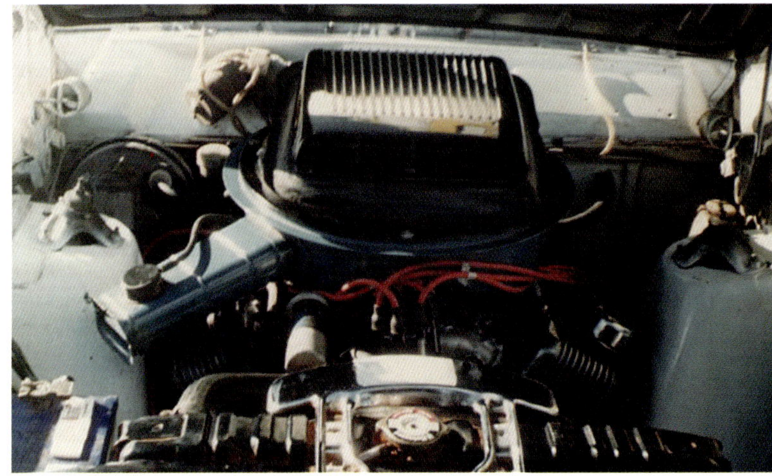

NY7246

52

FATSKD

Ford
RACING

FATSKD

460
HIGH PERFORMANCE

BRENDON'S
XY WAGON

The second-generation Falcons will always remain synonymous with performance. Indeed, when Ford Australia launched the XE model in late 1966, it included – for the first time – a factory V8 option. It then went a step further by building the very first Falcon GT, a model designed purely to win the increasingly important Bathurst endurance race.

With each new second-generation facelift came more and more outlandish performance models. The XR GT with its 289 V8 was replaced by the XT GT, and its 302. Then came the XW, with the bigger-still 351. While there was a GT option with the XW, it was the GTHO Phase I, released in August 1969, that became the Bathurst weapon. Then followed the XW GTHO Phase II, launched in August 1970; finally, the ultimate Australian Falcon – the GTHO Phase III, based on the latest XY model – was thrust upon its rivals in September 1971. The Phase III remains, to this day, the most legendary Falcon of them all.

So it comes as no surprise that owners of classic Falcons covet the second-generation cars, in all their forms. That includes the station wagons.

This 1971 XY (at some point fitted with an XW nose) Falcon 500 was factory equipped with an inoffensive inline-6. Its original purpose would have been to serve as a family runabout, or perhaps as the workplace of a travelling sales person. But this car's current owner, Brendon, has given it a whole new lease on life.

The theme for Brendon's build was 'Big is Best'. Indeed, the weedy

6-cylinder was pulled, and replaced by a 460 cu. inch big block stroked to an epic 545 cu. inch. As if that wasn't enough, a tunnel-ram sits atop the monster motor, further amplifying its sheer mass. The big block is backed by a C6 transmission with B&M Shifter, and Ford 9-inch differential.

All that power needs help to stop, so Brendon has installed Willwood 4-piston brakes in all four corners. Custom Cobra suspension ensures the fat kidney-bean mags and white-letter tyres are kept planted on the road. GT Falcon spotlights and front chin spoiler complete the look. It's a timeless package.

FATSKD

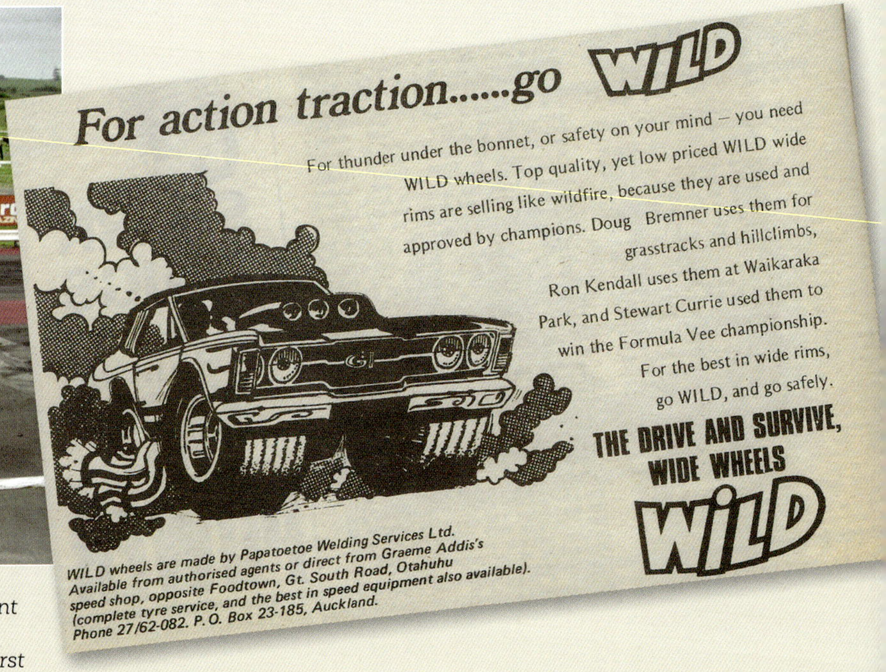

Top left: Cool-looking XY Falcon GT riding on Keystone mags, pictured at a drag racing event held on the Waiouru military airstrip in the early 1970s.
Top right: Brian Bowater's XY GTHO Phase III. This was the car raced by John Goss at Bathurst in 1971 before it was brought to New Zealand. Once its circuit racing career was over, it became a street car, fitted with slot mags and lake pipes, and a regular on the quarter mile.
Above: A gathering of Falcon GTs at Meremere drag strip in the 1990s.

Top middle: Former Grady Thomson GTHO Phase III, owned by father and son collectors Ross and Dallas.
Top right: The Parkes family XY Falcon with caravan in tow. Big Aussie sedans hauling caravans were a regular sight on New Zealand roads in the 1970s and '80s.
Above: Danny Clarke's XY GT pictured at Bay Park in 1982.

351 GT
SHAKER
BFGoodrich

youthful looking Dick Johnson was a special guest at a Southland Ford Falcon lub event.

RICK

Rick was a founding member of the Southland Ford Falcon Club. Four decades ago, the very first club run was held. Back then, classic Falcons were a much greater part of daily life, when they didn't fetch the high prices they do now, and most were used as everyday transport. Nonetheless, even then there was an appreciation for them, which naturally drew like-minded owners together.

Fortunately, Rick had the foresight to snap photos of the various Falcons in his life – and those of other club members – who were also close friends. Some of those images can be seen here.

MARINE PARADE
351 GT

LET THE
GoodTime's
ROLL
OFFICIAL MAGAZINE OF THE
SOUTHLAND
FALCON CLUB
SOUTHLAND
Ford
FALCON CLUB

Jim Richards in the Team McMillan XA Falcon GT heads the HQ Monaro of John Scott at Pukekohe.

CASTROL GTX SERIES

Production car racing enjoyed enormous success in New Zealand during the 1960s. Domestically, there was a sizeable local assembly industry, and naturally, people were going to want to race the products on offer. The same was true across the ditch in Australia. Indeed, Australian touring car racing focused largely on the fire-breathing Improved Production cars that contested the Australian Touring Car Championship in the 1960s, but the annual endurance contest for Series production cars at the car-destroying Mount Panorama road circuit on the outskirts of the sleepy town of Bathurst was fast gaining interest.

The Bathurst encounter required that cars remain truly stock-standard; just as they rolled off the assembly line. The race was held over a distance of 500 miles (800 kilometres), and just reaching the finish was an accomplishment in itself. But to win! Well, that offered manufacturer bragging rights. To say your car was both faster and more durable than those of your competitors truly carried some clout. And inevitably, greater commitment was required each year in order to emerge victorious.

Cars like the Ford Falcon GT; the GTHO Phase I, II, and III; the Holden Monaro GTS327 and GTS350; the Torana XU-1; and the Valiant Charger E-38 and E-49 all owe their existence to their manufacturers' desire to win at The Mountain under the strict Series Production rules.

By the mid-1960s, production racing was also gathering momentum in New Zealand. The annual Wills 6-Hour, which evolved

Top: Spencer 'Spinner' Black, living up to his nickname at Bay Park.
Above: One of the great GTHO competitors during the first season of the Castrol GTX series was Richard Brocklehurst in the John Andrew Ford supported Phase III.

into the Benson & Hedges 500, was to New Zealand what Bathurst was to Australia. But if the B&H 500 had a downside, it was that entries were limited to New Zealand-assembled cars. Therefore, the supercars being built in Australia to contest Bathurst weren't eligible.

Regardless, the Bathurst specials were still raced with enthusiasm in New Zealand, beginning in the late 1960s. There was no championship, but rather loosely assembled grids at various summer racing events for 'Production Cars'. Monaros and GT and GTHO Falcons were surprisingly common in New Zealand. Jim Richards, as a rising star of New Zealand motorsport, owned an HT Monaro GTS350. It was his daily transport, and it towed the Willment Racing Escort twin-cam with which he contested the New Zealand Saloon Car Championship. But the enthusiastic youngster wanted to get as much bang-for-buck from his travels as he could. So once the Escort was prepped and ready to race, he'd stick a number on the door of the Monaro, and contest the Production Car races.

As interest grew in production racing, particularly in the immense Bathurst specials being built in Australia, it was inevitable that a recognised New Zealand championship would be established for these cars. In the winter of 1971, the Motor Sport Association of New Zealand (MANZ) announced that a new championship would kick off in the upcoming season, to be held at several of the high-profile international events, with sponsorship from Castrol Oils and its GTX brand.

The 1971/72 Castrol GTX Championship was graced by the very latest and best Bathurst super cars, including Falcon GTHO Phase IIIs, Torana XU-1s, and Charger E-38s. The Chargers included three entries from New Zealand Chrysler assemblers Todd Motors, for Leo Leonard, Rod Coppins, and Graeme Lawrence. The GTHOs all had Ford dealer support, including the McMillan Ford example for Jim Richards, the John Andrew Ford car of Richard Brocklehurst, plus Robbie Francevic's Clayton Motors car.

Richards sprang away to an early lead in the championship, and built a healthy buffer. The big Ford's main rivals were the nimble Toranas. The Chargers took a little while to get on the pace, but when they did, they were well suited to the tight New Zealand circuits. Leonard in particular came on strong towards the end of the season, but it was all too late. Richards had scampered away to an unassailable lead, and wasn't to be caught. Tight little tracks like Levin's couldn't be more different from the steep mountain climb and long straights of Bathurst that the GTHO was designed for, but Richards made it work. He was crowned inaugural Castrol GTX Champion.

The new Charger E-49 was available for the 1972/73 championship, and several drivers switched to these. The big Falcons struggled to get their considerable power to the ground on the skinny little tyres they were using, whereas the Chargers were almost perfectly suited to the Kiwi tracks. The 1973 championship was won by Leonard.

The regulations were changed for the 1973/74 championship, largely because of the 'super scare' that had taken place in Australia, which brought production of the wild Bathurst specials to a screaming halt. In terms of the new rules, the 1973/74 championship required a larger production number of each vehicle in order to be eligible. The Ford teams, therefore, switched to the new Falcon XA GT, in four-door guise, while the Holden teams went to the HQ Monaro GTS350. Chrysler teams were forced to race the New Zealand-assembled Charger 770, which wasn't a match for its rivals.

Other than Neville Crichton's Monaro, the entire front-running fleet in the 1973/74 championship were XA GT Falcons. These included examples for Richards, Francevic, Inky Tulloch, Wayne Murdoch, Ron Rutherford, and Tony Warren. A quirky new points system would decide the championship. Each Island would have its own series, and the top cars would then qualify to contest two races: one in the North, and one in the South. With three different classes, the Champion elect would be the driver who scored the most points in his class. Somehow, what nobody appears to have realised, was that if two drivers from each class scored a first and second place in the two final races, there could potentially be six champions! In the end, it was Fiat 124 pilot Kevin McNamara who was crowned champion, while Wayne Murdoch won Group 1 for the big-engined cars.

Jim Richards emerged as the inaugural Castrol GTX Champion, winning the series in the handsome McMillan Ford GTHO Phase III.

The following year, it was Richards who scored his second Castrol GTX Championship in four years, when he guided his McMillan Ford XA GT to victory.

Not satisfied with leaving well enough alone, the organisers changed the regulations yet again for the 1975/76 Castrol GTX Championship. For the first time, American pony cars were eligible as the series adopted FIA Group 1 rules, and suddenly the GTX grid was awash with second-generation Z28 Camaros. Furthermore, whereas the previous incarnations of the GTX rules forbade any modification, the new Group 1 rules allowed a number of freedoms that ramped up costs and cut lap times. And as a result, the previously front-running Australian cars were made largely uncompetitive.

Aussie muscle did, however, get to insert a final full-stop on the GTX Championship. After two years under Group 1 rules and being dominated by the big-dollar Camaros, the series effectively priced itself out of the market. A consequence was that car numbers took a tumble, and interest waned. The 1977/78 Castrol GTX Championship, therefore, would be the last. It was won not by a Camaro, but by the beautiful lime-green XA GT hardtop of Barry Lloyd. Lloyd was a consistent top-three finisher throughout the season, and deservedly took the honours.

In all, the Castrol GTX series survived just seven seasons, but it rode the crest of a wave that was prevalent throughout the 1970s for high-performance production racers. And Falcons took top honours in four of those seven seasons.

Top right: The great Ernie Sprague sits in the middle of the front row at Teretonga in his GTHO Phase II.
Above: Barry Lloyd won the last Castrol GTX Championship in his superb XA GT hardtop.

Top: Alan's genuine Falcon GT, finished in Phantom Purple.

Top left: A young George Sheweiry purchased the famous Sidchrome Mustang when its Australian duties were completed, and used his GTHO Phase III to tow it to events. The GTHO also saw plenty of track duty.
Top right: XA Falcon hardtop forming part of the hugely impressive Stewart's Classic Cars collection.
Above: Michael Eden has owned and raced this XY Falcon for well over two decades.

351
351
GT
TOW

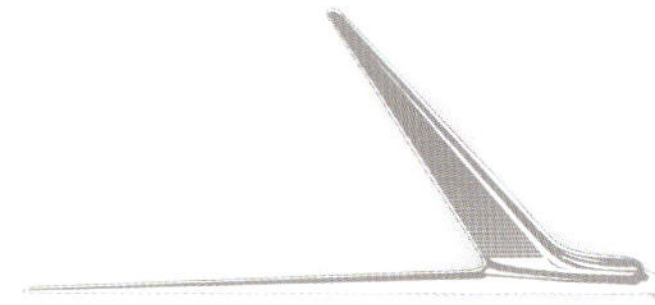

DARRYL'S
XA GT

Ford Australia offered a Falcon two-door hardtop for the first and only time with its third-generation range. Launched in 1972, the third-generation Falcon was the first to finally break away from the close resemblance to the American variant. Indeed, when the third-generation Falcon came to market, its American counterpart no longer existed. This was the very first truly Australian Falcon.

In 1968, Holden stole Ford's thunder when it released its two-door HK Monaro. Although Ford Australia had been producing hot Falcons since 1966, these commandeered the same body shell as that used for 6-cylinder police cars, taxis, and other mundane offerings. The Monaro, by contrast, boasted a unique body shape that reeked of sporty intentions, even with a wheezy 161 cu. inch inline-6 under the bonnet. The Monaro was never a big seller, but it gave Holden bragging rights, and provided it with a halo car.

When Holden launched the new HQ model in 1971, it again offered a two-door Monaro coupé. But now Ford was ready to counter with its own two-door, as part of the new XA line-up. Although Falcon buyers could order a two-door hardtop in the old XM and XP models from 1964 through 1966, these never had a V8 engine to go with them – unlike their American equivalent. When finally – in 1966 – Ford did deploy a V8 in its arsenal for the new XR Falcon, there was no longer a two-door hardtop body. Again, American Falcon buyers had the option of a V8 two-door. But with a much smaller market to work with, Ford Australia had to appeal to the masses, and forego the expensive tooling required

to produce the unique two-door. For the XA, however, it dug deep and produced perhaps the most striking vehicle ever to carry the Falcon nameplate.

The XA Falcon hardtop was unique to Australia, as was the XA itself. But clearly, its styling was heavily influenced by that of the larger American Torino, including the new two-door model with its near-horizontal rear roof line. The hardtop could be ordered with any engine in Ford's line-up, which topped out at the impressive GT, powered by the big 351 cu. inch Cleveland V8.

The XA Falcon GT was available in both four-door sedan and two-door hardtop form. However, in Australia, the two-door was the only body style to be raced. The Confederation of Australian Motor Sport (CAMS) introduced a new set of regulations, called Group C, for contesting both the Australian Touring Car Championship (ATCC) and the annual Bathurst endurance race. And for the period from 1967 to 1984, when the Falcon raced in a production-based guise, the XA was the only model to win Bathurst more than once. Allan Moffat and Ian Geoghegan won in 1973, while John Goss and Kevin Bartlett won in 1974.

While the Falcon hardtop was lauded for its superb styling, it was never a big seller. Australian customers tended towards a more conservative approach to car buying, and while the hardtop was a good-looking car, the four-door offered the same mechanical specification, but in a much more convenient package. And when Ford unveiled the new XD Falcon in 1979, a two-door body wasn't in the offering.

So the XA – XC Falcon in the two-door hardtop body shape was always rare, and even more so in New Zealand. But even so, old used-up examples could be bought for chump-change, and many were wrecked. The survival rate is relatively poor, which makes them even rarer today – and rarity, in many cases, equates to desirability. These big birds are highly desirable in today's classic car market.

This particular example is one of the most desirable of all the third-generation Falcons. It's a genuine XA GT two-door hardtop. It has been a racing car for most of its life, and despite its rarity, will continue to be raced. Its long-time owner, Darryl, had always loved the third-generation GT Falcons. Indeed, his father owned a Yellow Fire XA GT four-door sedan, which Darryl sometimes got to drive to school and

back. In the 1980s, Darryl ran a wrecking yard when this car came into his possession. It had rear-end damage, and he paid just $3000 for it. The factory colour was Red Pepper.

Darryl repaired the XA hardtop and began using it as a street car, but then started racing it at classic events. Indeed, its first race meeting was at Ohakea. Darryl was then invited to compete in one of the support races at the Wellington Street Race in the early 1990s. It was here, while at full-noise down Cable Street straight, that he went to stab the middle pedal and it went straight to the floor. With no brakes at all, he exploded through a large tyre bundle at the end of the straight. But even that barely slowed the heavy Ford down, so Darryl banged the gearbox from fourth into second gear, and put the Falcon into a spin. Although the speed diminished somewhat, he still slammed into a shipping container with enough force to cause damage to every panel other than the boot lid.

This was the second brake failure the Falcon had experienced. The first was at Manfeild, resulting in a gentle roll-over at the first corner. So Darryl decided to undertake a complete rebuild, which took six years and during which the Falcon became more of a dedicated racer. It was also repainted in Yellow Fire, the same colour his father's old XA GT sedan had been.

Darryl has owned the Falcon for well over thirty years and still enjoys getting it out from time to time. His second cousin, Peter Jacobs, has more day-to-day involvement with it. Peter has raced the Falcon regularly, both in classic events, and, for a while, with the Central Muscle Cars group. It has undergone several mechanical upgrades, the latest of which has made it a true fire-breather. It's now sporting a newly built 412 cu. inch stroker engine with alloy heads, which produced 800 horsepower on the dyno, although it's been pulled back to a more user-friendly 750. It runs a dry-sump, with Jerico NASCAR gearbox, and AP Racing 6-piston calliper brakes in the front, and 4-pistons in the rear.

The Falcon was always built for fun. Sure, it's now a valuable and desirable classic car, with rarity to match. But neither Darryl nor Peter see it that way. It's been in their lives since the 1980s, back when such vehicles could be had for just $3000.

CAUTION FAN MAY
START WITHOUT WARNING
DAVIES CRAIG PA

GT
TOW
creative signs

AUTOSPORT

MSD

Top right: Ben Gatt's epic XA Falcon hardtop drag car and transporter.
Above: XA versus XY at Meremere.
Left middle: Steve Twist's XA Falcon sedan, rolling on a period set of Keystone mags.
Left bottom: Stewart Quertier ordered a brand new XB GT sedan in Red Pepper in the early 1970s, but before it arrived, the dealership bumped up the price by another $1000. So he found this low-mileage XA GT in Jewel Green instead.

Top left, top right: Tony Johnson purchased this XA Fairmont in 1979, at age 18. Tony put the factory 302 car through its paces at every opportunity – and at every Give Way sign – and eventually treated it to a new custom paint scheme over the factory metallic blue.
Above left: After having fun with his XA Fairmont, Tony Johnson eventually upgraded to a factory XA GT.

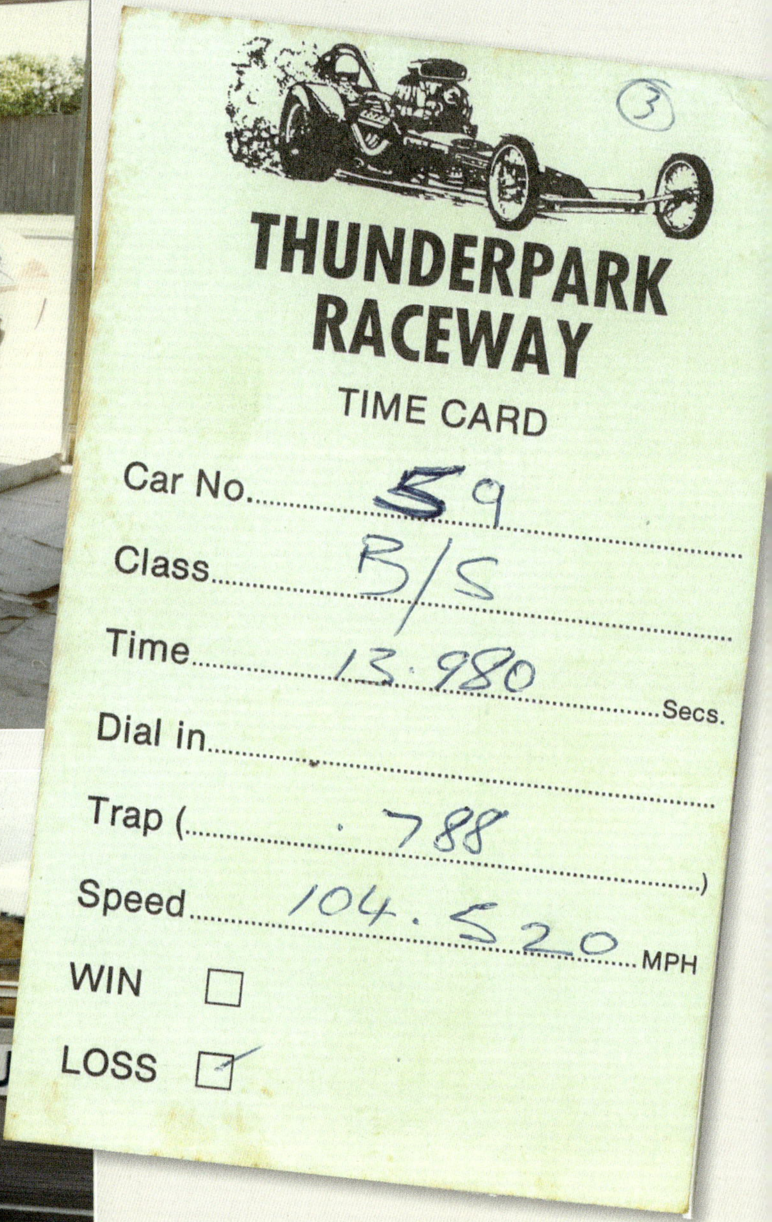

XA Falcon GTs are rare and valuable cars now, but back in the 1980s when Tony Johnson purchased this car, they were a common sight on Kiwi roads. Invariably, they weren't cosseted, and this one doubled as a daily driver and a part-time drag racer.

John McKechnie built this impressive XA Falcon Historic Muscle Car racer from a rusty shell.

YOKOHAMA
YOKOHAMA
YOKOHAMA
DRIVER
92
YOKOHAMA
YOKOHAMA

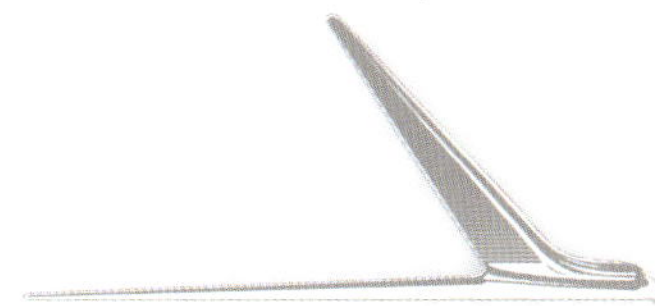

JOHN'S XA GT

We all have that story of the one that got away: the car we should never have sold. As for this Falcon – it got away twice, from two different owners.

This particular Falcon XA GT hardtop appears earlier in this book, when owned by Tony Johnson. Tony owned the car for a number of years in the early to mid-1980s, having purchased it in Whanganui, and then sold it to start his own business. Before selling, he consulted a couple of his buddies as to his sanity in doing so. He loved the car, but was assured, 'It's only a Falcon. It doesn't have a worldwide value. It'll never be worth that much and you can always buy it back again one day.' So off it went.

In 1986, it was purchased by John Pope from a dealership in Auckland. John traded a Mercury Cougar for it, and although he knew it'd been drag-raced in a previous life (by Tony Johnson), John planned to circuit race the Falcon anyway.

Back in the 1980s, Australian muscle cars were affordable and plentiful, and made good racing fodder. Certainly, they were a cheaper option than a Mustang or a Camaro, and parts were still available at good prices.

John had his first racing event at Pukekohe. The Falcon hit the track in much the same state it was in when he bought it. It was still road registered, had the very same engine, and little in the way of safety equipment. John even retained the tyres. But like all racing cars, it evolved, and became more purposeful.

Eventually, John decided to upgrade the Falcon with an XC nose, so he could also fit the improved front and rear XC spoilers, plus the bonnet scoop. An XC panel van was the source of the required parts. The XA GT front was sold. For classic racing events, he had to run an XB nose as the later XC one wasn't allowed.

John raced the Falcon for a number of years, competing in the Production Muscle Car class, and at various classic events. In 1992, in a classic race at Whenuapai, the Falcon sustained a huge hit on the driver's side when impaled by an out-of-control Porsche. The damage was significant, and required a major rebuild. Fortunately, fellow Falcon hardtop racer Craig Pullman had recently convinced John he should replace the existing aluminium roll cage, which was basic (at best), with a much more substantial cage, including side-intrusion bars. That decision saved John from sustaining serious injuries.

While John came away virtually unscathed, the same couldn't be said for the Falcon, which required a replacement driver's door and extensive repairs to the rear quarter panel. Another panel van gave up its door and door frame, while the rear quarter was grafted back in place, patched, and fitted with the wheel arch from a front mudguard.

After the Falcon had been rebuilt, John decided to sell it. In the early 1990s, these cars weren't coveted to anywhere near the extent they are today, and amazingly, John struggled to find a buyer. Eventually, he decided to part it out. The body shell went to someone in the central North Island who planned to put it back on the road. By now, it had two engines, including the original GT unit, and a dedicated race motor. The former ended up in a T Bucket hot rod, while John swapped the race motor for a Datsun 1200 racing car. The motor ended up in a speedway Saloon. Such were the perils of trying to sell an old Aussie race car thirty years ago.

Fast-forward fifteen years, and John decided to try to track down his old car with the view to buying it back again. He contacted the Post Office who informed him the car has not been registered since 1994. He has since tried again, but continues to draw a blank. Despite various attempts to trace the Falcon, the trail continues to go cold. The fire still burns, however, and the search continues. Currently, all John has of the Falcon is the crushed driver's door from the Whenuapai shunt.

Interestingly, John has since made contact with the Falcon's former owner, Tony Johnson, who also suffers from seller's remorse. This very special XA GT has yet to resurface, but it has left a lasting impression on two of its owners.

Incidentally, the XA GT isn't the only car in this book to appear twice with two different owners. John also owned an XC Falcon Cobra (number 121 of 400 made), with which he towed the XA race car to events. The Cobra has been in the long-term ownership of Darryl, whom we've met elsewhere on these pages.

FERGUSON
22
JOHN POPE
2
SMITH

At one time, John owned this genuine Falcon XC Cobra, which doubled as the tow-car for the Falcon race car. This same XC Cobra features elsewhere in this book, owned for many years by Darryl.

Secure-a-caR
2
Segedins
OLYMPUS
YOKOHAMA

SHERMAN
GY 7611

SMITH

Gary Raiti's beautiful XA Falcon 500 race car heads Tony Mexted's 1970 Boss 302 Mustang at Hampton Downs.

Gary Cawthra's stunning black XA Falcon hardtop drag car assaults the eardrums of race-goers at Masterton Motorplex. Gary has owned the Falcon for more than two decades, and it has evolved massively in his hands.

ENGINE

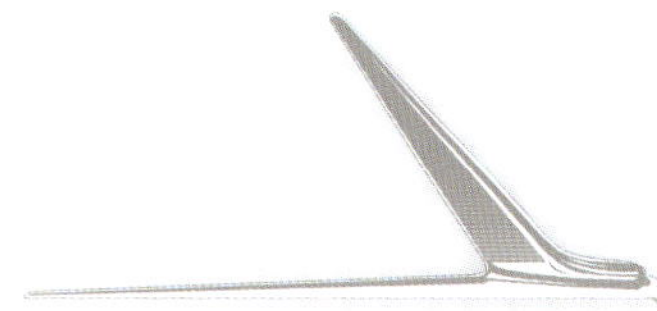

HAYDEN'S

XA & XB

Falcons have been popular fodder among hot rodders and customisers virtually from day one. And the third-generation XA – XC cars are perhaps the most versatile of them all.

This pair of third-generation models are owned by Hayden, and are modified to the hilt. Indeed, they look like life-sized Hot Wheels cars. They are a 1973 XA two-door hardtop and a 1976 XB sedan.

Of the two, the XB is the cruiser. It runs a 351 cu. inch with tunnel ram, and rides on US Rambler wheels, measuring 18 x 9 inches in the front, and a whopping 20 x 10 in the back.

While the XB isn't most people's idea of mild, compared with the XA hardtop, it really is. The XA is powered by a 358 cu. inch topped with 671 supercharger, and a pair of Demon 750 carburettors. Backing the 358 is a Powerglide transmission and Ford 9-inch differential. Again, the wheels are big: 20 x 10 inches, front and rear. The door handles are shaved, and the car wears a pearl-white wrap.

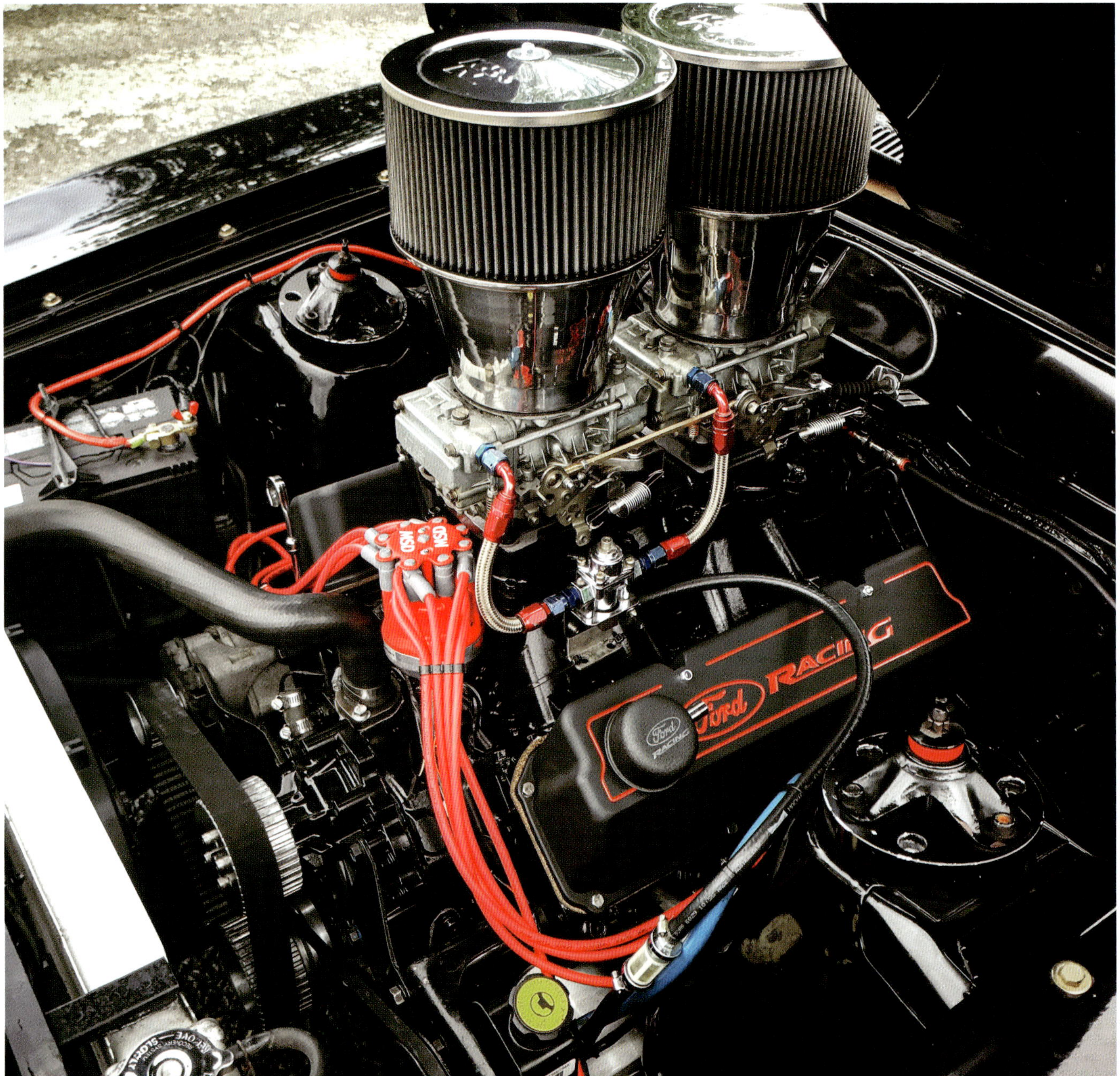
MSD
Ford

DANGER
BUDDAS

ENGINE
MAIHEM

Ford RACING
MSD

The WASP
GT351

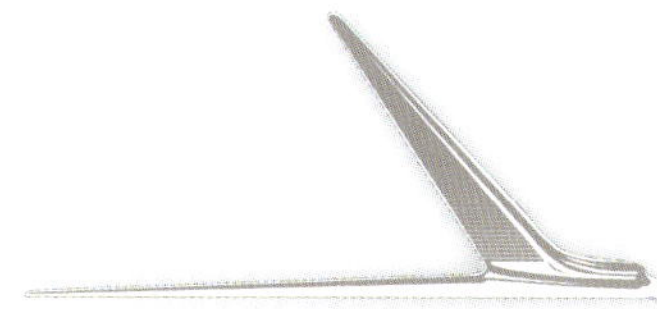

LANCE'S XB GT

A more interesting Falcon you will not find. It's unique! To start with, its factory-applied modifications set it apart; and then, this 1975 Falcon XB GT has remained in the same (extended) family since it was just over twelve months old. Furthermore, it has travelled a grand total of just 64,000 km from new.

Elton Harrison, of Levin, was a caravan dealer, and owned Pioneer Caravans. In 1975 he visited his local Ford dealer (now Courtesy Ford), and optioned up this unique Yellow Blaze GT. He wanted a mix of performance and style, and chose a 4-speed manual, with manual steering, along with the high-performance 351 engine. He also ordered electric windows, a sunroof, and a set of 14-inch Magnum 5-slot mag wheels.

The most dramatic option, however, is the bold Continental vinyl roof. Falcon buyers had the choice between full or half vinyl; Elton chose the latter. The Continental roof had also been an option on the XA Falcon, both in sedan and in coupé form, and it differed from that fitted to the luxury Landau coupé, with the vinyl wrapping back around and beneath the side windows. There were even choices on how far the lower vinyl protruded forward.

Harrison then had a set of custom 'The Wasp' decals made for the front mudguards and the air cleaner, to underline its wild colour scheme.

Harrison kept the Falcon for just over a year, before 16-year-old Darryl Lucas spotted it on the lot at Peter Hogg Motors in Palmerston North

with around 12,000 kilometres on the clock. Darryl's father, Herb, owned a Falcon XA GT at the time, which Darryl sometimes drove to school! He convinced Herb the bright yellow XB would look good in the Lucas family driveway and, to that end, it went home with them.

Herb kept the XB for many years. He had it repainted in the 1980s, purely because the factory paint was never that good, but other than that it remained largely as Elton Harrison had ordered it back in 1975, albeit that the Wasp decals were changed slightly. It was driven infrequently, as Herb moved on to more modern and comfortable daily drives.

Herb's son Darryl, clearly influenced by the XA and XB Falcon GTs of his youth, has owned an XA GT hardtop and XC Cobra (see separate chapters) for many years. But he wasn't the only family member who caught the Falcon bug from Herb. Lance, Herb's nephew, grew up around the Yellow Blaze XB, and although Herb had no plans to sell the car, in the late 1990s he agreed to accept an offer from Lance, perhaps doing so because it meant the car effectively remained in the family.

Under Lance's ownership, the Falcon was enjoyed and properly exercised, but never mollycoddled. It was, however, driven only on occasion, and for this reason it has still rolled over fewer kilometres in nearly half a century than many modern vehicles achieve in a year.

Sadly, early in 2020 Lance passed away suddenly. The Falcon, however, remains in the family, and will continue to do so. One of Lance's sons, Richard, drove the Falcon to his funeral, while Lance's brother Peter drove it home. 'Richard drove the Falcon to the funeral,' explains Peter. 'Then I drove it home and did a skid just to piss him off because, he never drove it like that! Then Richard drove it to the unveiling of his father's memorial.

'His sons Richard and Blair have taken it out with their boys for a couple of cruises just like Poppa did with "his" boys.

'Jill (Lance's wife) told me not so long ago that she was always his second love; his first being the XB. Lance was a real Ford nut and as far as he was concerned there was nothing else that compared.

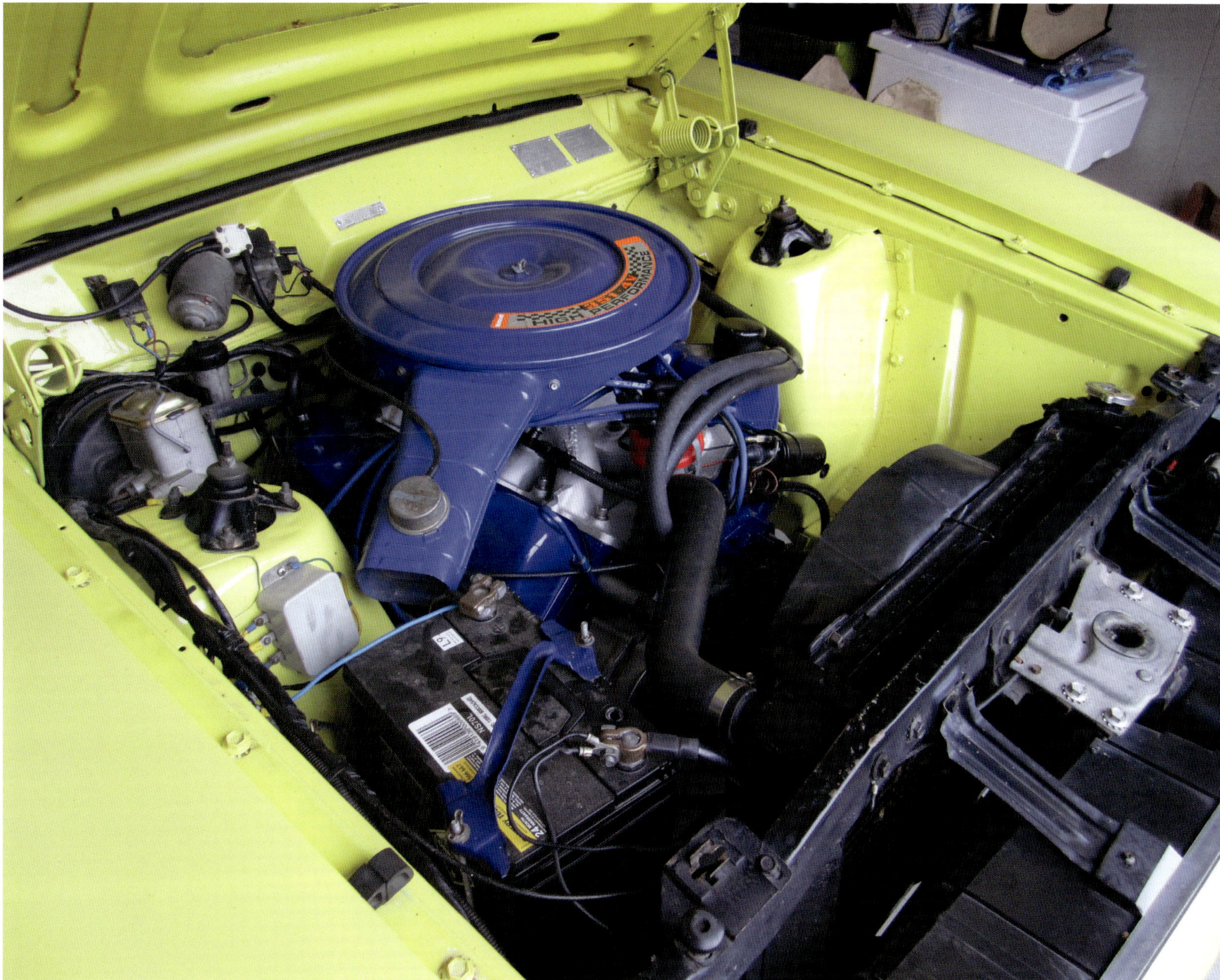
351 4V
HIGH PERFORMANCE

The WASP

GT351

GT351

Top right: Steve Twist's XC.
Above: Craig Pullman's XB Falcon hardtop muscle car racer. He later rebuilt it as a wild Sports Sedan.

Top right: Falcon XB course car follows the big field away at the start of the annual B&H 500 at Pukekohe.
Bottom right: The Robbie Francevic/Alan Woolf XB Fairmont V8 during the 1974 B&H 500. Only New Zealand-assembled cars were allowed to compete in the B&H race.
Above: Gary Bromley's XB GT circuit racer.

PROWEAR
SPECIALISTS IN CUSTOM RACESUITS, CREW SHIRTS AND TEAM GEAR!
CUSTOM SHIRTS NZ
WWW.CUSTOMSHIRTS.CO.NZ
AEROFLOW
PERFORMANCE PRODUCTS
1 AUTO ONE
AUTO PARTS & ACCESSORIES
RIGHT GEAR! RIGHT ADVICE!
0800 288 666
www.autoone.co.nz
NEIL ALLPORT
MOTORSPORTS
Your motorsport accessories, parts & service provider
SAMCO sport
Aeroquip
Setrab
Oil Coolers
sparco
ATL
RACING FUEL CELLS
34 Olive Road, Penrose, Auckland, New Zealand
09 579 0113 neilallportmotorsport.nz
AUTO TRANS
AUCKLAND
AUTOMATIC TRANSMISSION SPECIALISTS
0800 686 999
TOAST HUT
Beef & Gravy Rolls
COUNTIES
Hoosier

GARY BOGAART

Gary Bogaart is best known for his commitment to drag racing, both on and off the track. As well as being the current Track Manager at Meremere Dragway, he has raced a number of tough cars, the most prominent of which have been Falcons.

Gary is a National Record holder and race car builder. Among his creations was the 2007 BF Falcon Top Doorslammer. This tube-frame monster was commissioned by Anthony 'Dutchie' Wijdeven, with the intention that Gary would drive it. After several years, and countless achievements, eventually Dutchie took the helm himself.

For many, however, it's Gary's bright orange XB hardtop that remains a favourite. He has run a variety of engine combinations in the Falcon over the years, the most recent of which is a 408 cu. inch small block Ford on nitrous. It's famous for its stout wheel-stands, but with a set of wheelie-bars installed, the Falcon runs some impressive passes. Despite this being a street car, Gary's Falcon has clocked 8-second ¼-mile times at 160 mph. The Falcon has also been raced by Gary's teenage son Nick in recent years.

In addition to the drag cars, Gary has also owned a stream of fleet-footed Falcon street cars, many of which have also seen track duty. They range from XDs through to sixth-generation XR8 and long-wheelbase Fairlane.

SVO 58

Rylock
PERFORMANCE CAR

BRIAN ROBERTS LTD
TOWING & PANEL
RACING
¼ MILE

AEROFLOW
aeroflowperformance.com
THE ROCK
NEIL ALLPORT
AUTO TRANS
TOOL ORGANISATION AT ITS BEST!
Rylock
0800 4 RYLOCK
www.rylock.co.nz
PERFORMANCE CAR
Harley-Davidson
1/3 RPM
RISLONE
MOUNT SHOP
1/4 MILE
MEREMERE DRAGWAY

Start of the 1978 B&H 500 at Pukekohe for New Zealand-assembled production cars.

George Sheweiry's Falcon XC Cobra, pictured at Bay Park.

MAG & TURBO
TYRE & SERVICE CENTRE
LOWERHUTT
BURGERFUEL
ENGINEERING THE ULTIMATE BURGER
CLIFF SCOTT
ENGINEERING
SIGNBIZ
VALLEY VALET
Professional Vehicle Grooming
PLUMB 2 PLEASE
VEHICLE SERVICE
Action
CIVIL LIMITED
O'SULLIVAN
EXIT
BlackHawk
TAXI Co.
IQ6816
CAUTION

Marty's

XC

In recent years, originality in classic cars has become king. Enthusiasts want to see what vehicles looked like when they rolled off the assembly line, warts and all, and that originality is lost forever when a car gets restored or modified. In the case of many older cars, restoration is a necessity. Others, however, have been fortunate to have had caring owners who've preserved them, through the highs and lows.

Finding an original classic car that is still in the care either of its original owner, or of its original family, is ultra-rare. But that applies to this superb 1979 XC.

This car is owned by Marty. His grandfather ordered it new in late 1978 from Metro Ford, Porirua, when he retired as Police Commissioner. It was first registered in January 1979. Marty's grandfather ordered it with passenger door mirror, sunroof, GS alloy wheels, auto reverse tape deck, electric aerial, and electric boot release. Since it rolled off the Seaview assembly line over 40 years ago, it has travelled just over 102,000 kilometres.

When Marty's grandfather was finished with the Fairmont, he passed it on to Marty's parents, and in 2000, it went to Marty. Since taking ownership, Marty has added a few careful upgrades, to make the Fairmont more user-friendly. These include a set of 17 x 8-inch Cragar SS1 wheels, rear seat belts, upgraded sway bars front and rear (a rear sway bar wasn't a factory option), and Black Diamond 750CFM Quick

Fuel carburettor in place of the factory Carter. Bosch ignition with MSD is next on the list. However, everything Marty has done – other than the addition of the rear sway bar – can be reversed, and all the original parts are kept safe in storage.

Marty's grandfather passed away in 1991. He'd no doubt be thrilled to know his beautiful Fairmont remains a cosseted family member.

IQ 6816

Winfield 25's
Shell

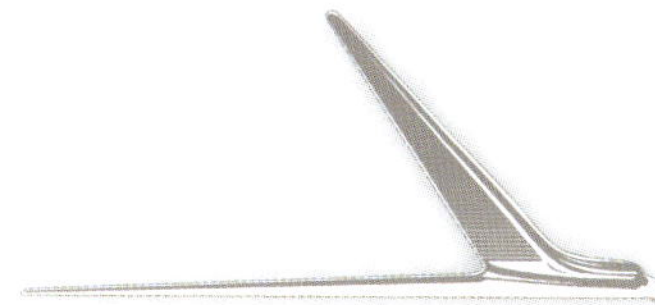

JIM RICHARDS' FALCON SPORTS SEDAN

Interestingly, perhaps the greatest racing Falcon ever built in New Zealand was never actually intended to be raced in New Zealand.

By the mid-1970s, Jim Richards and his young family were settled in Australia, and Richards was already a household name in Australian motorsport. In 1975, he shipped his Sidchrome Mustang across the ditch to compete in the Australian Sports Sedan Championship, and associated big-dollar Sports Sedan competitions.

In the early and mid-1970s, Sports Sedan racing was the most popular motorsport category in Australia. Although touring car racing had overtaken Formula 5000 in terms of popularity, Sports Sedan racing was even bigger again. Sports Sedans had its roots in a category called Sports Racing Closed, which was a largely lawless division that was strongly supported at New South Wales circuits, Oran Park in particular.

In 1973, Sports Racing Closed regulations were given a polish, and the class was renamed Sports Sedans. The make-over was part of a larger operation, as the Series Production and Improved Production regulations also underwent major upheaval. Series Production cars were essentially stock-standard cars, the likes of which competed in the annual endurance race at Bathurst. Improved Production were highly modified production vehicles, which competed for the Australian Touring Car Championship. Series Production rules spawned low-volume performance specials, such as the GTHO Falcons, E-38 and E-49 Chargers, GTS325 and GTS350 Monaros, and the XU-1

Toranas. Improved Production attracted mostly American pony cars: Camaros and Mustangs.

1973 was also the year in which the Australian motorsport governing body, CAMS, introduced Group C, which would incorporate both the ATCC and Bathurst. Group C cars were similar to Series Production, but with a lot more freedoms allowed, so low-volume specials weren't required. Concurrently, CAMS also recognised Sports Racing Closed, which it rebranded Sports Sedans. This was where many of the Improved Production cars ended up.

The crowds loved the wild and innovative nature of Sports Sedan racing. There were rules, yes, but they embraced the concept of mad creative freedom. Cars could be fitted with any production-based engine, although those that didn't roll off the assembly line with a V8 were limited to a maximum engine size of 5000 cc. Factory V8 cars could go to 6000 cc. The engines could be moved up to 12 inches from their standard location. As a result, Bryan Thomson commissioned a VW 1500 fastback with a fuel-injected small block Chevrolet, mounted behind the driver. Aero-aids were also permitted.

In 1970, the Toby Lee Series ('Toby Lee' was a shirt brand owned by department store Grace Brothers) was organised at Oran Park. Originally it was for Series Production cars, these being of the same type that competed at Bathurst. But because of the crowd-pulling appeal of Sports Sedans, the Toby Lee Series switched to these, starting in 1973. For 1973, the prize fund was A$15,000. In 1974, it doubled, to $30,000 – a considerable sum at the time. For this reason, the Series enjoyed quality competition from the best Sports Sedans in Australia, with John Harvey eventually emerging as the 1973 Champion, driving Bob Jane's wild Holden Torana, fitted with an enlarged fuel-injected Repco V8 Formula 1 engine. In 1974, the series was won by John McCormack, driving a Valiant Charger fitted with a Repco-Holden Formula 5000 engine, which was positioned so far back in the car that it sat alongside the driver.

Spotting a lucrative opportunity, Melbourne businessman Bob Jane established his own Sports Sedan series, held at his Calder Park Raceway. Jane organised sponsorship from Marlboro cigarettes, to the tune of $100,000. While it was the cheap and scary creations that propelled Sports Sedan racing to stardom, all the interest and money pouring into the category generated a whole new breed of big-dollar machinery.

So when Jim Richards arrived in Australia with his Mustang in 1975, Sports Sedan racing was booming. Finally recognising that Sports Sedan racing was a significant part of the Australian motor-racing landscape, CAMS granted the category championship status, with the very first Australian Sports Sedan Championship (ASSC) held in 1976. Despite the Sidchrome Mustang lacking much of the bristling technology of the Australian cars, Jim Richards enjoyed immediate success. He won the 1976 Marlboro $100,000 Series and finished fourth in the 1976 ASSC.

But even in 1976, Richards knew the Mustang's days were numbered. Allan Moffat won the inaugural ASSC, switching between his DeKon Monza tube-frame IMSA car, and his quad-cam GAA V6 Capri. Moffat won because he'd built a healthy points buffer early in the season, winning three of the first four races. Tony Edmonson, now driving the ex-John McCormack Repco Charger, won the other. But Frank Gardner debuted his epic new Chev Corvair in Round 5 (of seven), and blitzed the competition; he clean-swept the last three races. He then went on a winning spree in 1977, comfortably claiming the Australian Sports Sedan Championship. Richards again finished fourth.

The Gardner Corvair was effectively a purpose-built tube-frame car set out like a Formula 5000 single seater, and using Formula 5000 componentry. It was epic. And nothing could get near it.

When he won the 1976 Marlboro series, Richards did so without winning any of the races. Victories fell the way of Moffat, Bob Jane (Monaro), and Gardner. Indeed, Gardner won the final three races, just as he had done in the ASSC. Richards won by being consistent. But to stay in the game, Richards needed a new car.

Richards leads Brent Bullivant out of the Bay Park hairpin in 1979. Bullivant built his remarkable V8-powered Capri as a teenager, and Richards was so impressed, he hired him to race-prepare his Falcon!

Even by 1976, construction of the Sidchrome Mustang's replacement was well under way. Just as he had done with the Mustang, Auckland car dealer Jerry Clayton funded the new project. It was to be a Ford Falcon hardtop, a model surprisingly overlooked by the Sports Sedan contingent, despite its popularity and success in Group C touring car racing. Ford Australia came to the party and supplied, free of charge, a brand new XB Falcon hardtop body shell, which was shipped to New Zealand, where Murray Bunn began construction. In addition, Ford also stamped a set of aluminium door panels, plus bonnet and boot lid.

Modified sedan racing evolved rapidly in the 1970s, and despite being conceived just four years after the Sidchrome Mustang, the new Falcon racer would be radically more advanced. From a quiet spot in the back of his Bunn and Cummins workshop, Murray Bunn constructed the new weapon. Richards and Bunn decided to stick with the blue oval, largely because of their experience working with the Ford powerplant in the Mustang.

Both Jim and Murray took a trip to the United States to visit racing legend Dan Gurney, at his All-American Racers (AAR) facility. It was there they purchased a stockpile of components for the Falcon project. AAR had a sizeable collection of parts from its various racing programmes, including some ultra-rare 3-valve Gurney-Weslake cylinder heads. These had been developed for the Can-Am sports car series when AAR was running Ford-supported cars, including a Lola T70 and McLaren M6B. AAR ran small block engines in Can-Am, which utilised a special aluminium engine block.

Although Jim and Murray hoped to purchase a set of the 3-valve heads, ultimately, the asking price was beyond their budget. However, they returned home having secured a Ford aluminium engine block, regular 2-valve Gurney-Weslake cylinder heads, Gurney-Eagle front uprights, and McLaren M6B rear uprights. From the AAR-sourced parts, Mick Webb built a 366-cu.-inch motor comprising the alloy block, Gurney-Weslake heads, and a mechanical fuel-injection system.

Bolted to the back of the motor was a 5-speed Hewland DG300

transaxle, originally designed for mid-engined single seaters and sports cars. The Sports Sedan regulations required a gearbox be mounted directly behind the engine, but didn't stipulate what type of gearbox it had to be. Meanwhile, the differential was a 2-speed Hewland LG500. This unit was designed for Indycar racing on oval speedways, where one gear was used for starting, and the other for racing. So, effectively, the Falcon had two diff ratios. Because the DG300 was a transaxle, and was never designed to transmit power to a differential, when an output shaft was mounted out the back of it, it turned in the wrong direction, so the LG500 was turned upside down and back to front.

The new Falcon shell was chopped up, with only the roof, rear quarters, rocker panels, and outer sections of the floor remaining. The rest was constructed from scratch. Bunn built the car on a chassis plate. He designed it with the rocker panels sitting just 3.5 inches from the ground, so the car would be very low. A tube-frame chassis was fabricated on the chassis plate, inside the gutted body shell. The engine was moved as far back towards the centre of the car as the regulations would allow, to the point where it sat inside the cabin, alongside the driver. Indeed, the front of the engine was in line with the firewall. The rear wheel openings were radiused out to accommodate the big 15 x 10 custom-built magnesium wheels. The entire nose section was drooped down to better slip through the air. Even though it started as an XB, it ultimately became an XC.

The plan was to have the Falcon ready for the start of the 1977 Australian Sports Sedan Championship. However, delays in its construction meant this didn't happen. Although Jerry Clayton owned the car, Richards himself would have to finance the racing programme, and while Sidchrome Tools was to sponsor the car, such was the extent of the delays in the building programme that the Sidchrome partnership ended before the Falcon ever saw the light of day. Sourcing of parts and getting clarification on some of the grey areas in the Sports Sedan regulations contributed to the setback. Meanwhile, Richards soldiered on throughout 1977 with the outdated Mustang. It ended the year wearing a bleak coat of white paint after the partnership with Sidchrome concluded partway through the year.

By the time the Falcon was finally completed in late 1977, Sports Sedan racing in Australia had taken a kicking, and lost some of its lustre. This was largely the result of the dominant Gardner Corvair, which simply wiped the floor with the opposition, to the point they largely fell away. Bob Jane, driving his constantly evolving HQ Monaro coupé was to be Gardner's only real opposition in 1977, but even the wealthy tyre tycoon had no answer to the low-slung rear-engined Chevy. Having won the ASSC, Gardner promptly retired as a driver to focus on team management. For 1978, popular Holden battler Allan Grice would step aboard the Corvair.

Richards flew to New Zealand in early 1978 to test the newly completed Falcon at Pukekohe. The sleek white bird was a work of art, and hit the ground running. Straight away, Richards was lapping at record speeds. The Falcon was then shipped to Australia where it underwent final preparation for the opening round of the 1978 Australian Sports Sedan Championship, at Surfers Paradise. It was during scrutineering that Frank Gardner got his first look at the new Kiwi challenger, and at which he drolly said to Richards, 'I think you're going to be a pain in the arse!'

Seven rounds made up the 1978 ASSC, and either Grice in the Gardner Corvair, or Richards in the new Falcon, won every race. Grice took maximum points in the opener at Surfers Paradise, while Richards returned serve in Round 2 at Adelaide. The Kiwi then continued his momentum in Round 3 at Wanneroo (now called Barbagallo Raceway), while Grice came out on top in Rounds 4 and 5. Richards won Round 6 at Calder Park, with Symmons Plains being the decider. Two heats were held at the little Tasmanian circuit, and Richards, from pole, was leading the opening encounter, with Grice chasing hard, when the Falcon suddenly slowed with a broken pushrod, gifting Grice the win. The motor repaired, Richards blitzed Grice in Heat 2, amassing a total of 48 championship points to Grice's 45. However, the regulations only took the best six finishing positions, which placed the pair on equal points. The championship was decided, therefore, by Grice having

Jim Richards heads Warwick Gray's Holden-powered Viva, John Briggs in the former Red Dawson DeKon Monza, and George Sheweiry in Richards' old Mustang.

scored four Round wins to Richards' three. The broken pushrod at Symmons Plains was ultimately the title decider.

Richards may have lost the championship, but in his new Falcon he truly had a weapon capable of downing the previously all-conquering Corvair, a feat nobody had been able to achieve since its first appearance in 1976.

In early 1979, Richards shipped the Falcon home to New Zealand, where it contested the 3-round Open Saloon Car series, held at Bay Park, Pukekohe, and Manfeild. Big-banger saloon car racing had met its end in 1977, when Motor Sport Association of New Zealand (MANZ) decreed the category unsustainable. Indeed, car numbers had plummeted as technology and costs spiralled. Although the South Island had the popular Open Saloon Car Association (OSCA) series, the fans mourned the loss of the earth-shattering V8s in the North. While the Open Saloon series carried no championship points, it nevertheless attracted the best and fastest sedans from around New Zealand, plus the Richards Falcon and the John Briggs tube-frame DeKon Chevrolet Monza from Australia. The Briggs machine was the one first imported to New Zealand from the US by Kiwi racing legend Red Dawson, and the same car that ended his illustrious racing career when he barrel-rolled it at Manfeild in late 1976.

Each of the three events had two heats, and each heat featured a handicap start, with the slowest cars in qualifying set off at timed intervals, working back to the fastest. As it transpired, the two lead protagonists were Richards and Leo Leonard, driving the wild tube-frame PDL Mustang II, which was now fitted with a McLaren Formula 5000 rear wing perched on spindly struts way up above the rear bodywork. Each race saw Richards and Leonard setting off together, last away, and slugging it out as they surged forward through the dense traffic. This was a spectacle for the ages. At each event, the pair traded fastest lap times, annihilating existing lap records, and leaving them in tatters for years to come.

Following its Kiwi encounters, the Falcon was shipped back to Australia where it underwent upgrades in preparation for another shot at the Sports Sedan Championship. Having been pushed so hard, Gardner ensured the Corvair was further developed, and it received, among other things, a new and more powerful Chevrolet V8 engine. But 1979 wasn't to be a fiery rematch of 1978. The Falcon had been run on a tight budget, and now the cracks were beginning to show. Components that had been new in 1978 needed to be replaced following the stresses of a hard-fought season. But Richards' finely stretched finances didn't allow for this, and for this reason, mechanical failures became the theme the 1979 ASSC. Indeed, for Richards, 1979 was a win-or-bust year. After blazing to victory in the opening round at Lakeside, he was struck by a series of frustrating failures that took him out of contention for race victories. Indeed, Lakeside was to provide the single highlight of the season for him, and he was left languishing in the points. Certainly, the Falcon possessed the potential to win the championship, but the sponsor that would afford the purchase of top new components was never forthcoming. Big M, a milk drink, was the main sponsor for 1978 and 1979, but its contribution was just A$1200 per event; a pittance to successfully campaign a car like this. Richards had to scrape together smaller sponsors, as well as appearance money, just to keep the car running.

After a catastrophic engine failure in Round 7 at Symmons Plains, the Falcon was parked for the remainder of 1979, while the motor was slowly rebuilt in preparation for the 1980 season. However, car owner Jerry Clayton decided he needed to sell the Falcon to expand his business empire. Surprisingly, it was one of Richards' on-track rivals, Bob Jane, who purchased the Falcon for Richards to drive. Jane owned a chain of tyre stores called Bob Jane T-Marts. Indeed, in 1979, Richards purchased his own franchise. Jane's gesture was incredibly generous, although Richards still had to fund the racing program himself, and again, component failures as a result of worn parts were to blunt his efforts. He scored two victories, and placed third in season points, despite not completing the full race schedule. At the end of the 1980 season, Jane decided to sell the car, and Richards stepped up to purchase it. He'd raced it for three seasons, but it had always been someone else's car. But 1981 proved an unsuccessful year in the Falcon.

Richards had catapulted to the top level of the sport by the late 1970s. He co-drove with Peter Brock in the Holden Dealer Team and the pair won the Bathurst 1000 in 1978, '79, and again in 1980. Indeed, his services were much in demand. In 1982, he joined his old rival Frank Gardner at the JPS cigarettes sponsored BMW team running 635 CSis in the Australian Touring Car Championship, and a wild 318i Turbo in the new Australian GT Championship, which replaced the old Sports Sedan Championship.

With Richards' racing career soaring, the Falcon Sports Sedan sat idle, until eventually it was sold to Tony Hubbard, less motor. Hubbard installed a fuel-injected small block Chevy V8 in the Falcon and ran it in the newly established Australian GT Championship. Paul Halstead, owner of The Toy Shop, purchased the Falcon engine for a wild De Tomaso Pantera driven by Kevin Bartlett in the GT Championship. Indeed, Richards himself won the opening two rounds of the championship in the Team JPS 318i Turbo, before getting caught up in a spectacular multi-car pile-up moments after the start of Round 3 at Adelaide, when the Peter Brock-driven Chevy Monza owned by Bob Jane snapped left and into the wall, before bouncing back into the path of the chasing pack. The JPS BMW was flipped onto its lid and all but destroyed. Ironically, Hubbard was also caught up in the incident in the old Richards Falcon, which also sustained heavy damage.

Hubbard rebuilt the Falcon and continued racing it in 1984. Despite running a limited campaign, he scored an excellent second-place finish in Round 3 at Calder Park. He continued to race the car during the next few seasons, and even ran it for a time fitted with a twin-turbocharged 253-cu.-inch Chevy, but its best years were behind it. Hubbard eventually sold the Falcon to Robin Doherty who campaigned it until a new regulation ended its front-line career. To keep Sports Sedans current, a new ruling decided the cars had to have a body shell no older than ten years old. To that end, Doherty stripped the running gear from the Falcon and installed it into a Holden Commodore.

In 1993, the Falcon was purchased by Graham Booth, who raced it initially in State Sports Sedan events, fitted with a NASCAR motor he owned. When this engine blew in 2004, he decided it was time to start returning the Falcon back to its original Jim Richards guise. He tracked down the Gurney-Weslake heads to a hot rodder who had them fitted to his T Bucket. All the running gear that ended up in the Doherty Commodore was also purchased back. Booth got moulds for the original wheels, and has cast a new set from aluminium. It'll be a long and expensive transformation, but Booth wants to ensure it's done properly.

That fleeting 1979 appearance was the only time this epic machine raced in New Zealand. This was a Kiwi-designed and Kiwi-built car, and it was owned and driven by a Kiwi. But it was not intended to be raced in New Zealand. Indeed, it was created to take on and beat the very best that Australia had to offer – and it very nearly toppled them all. Had Richards gained the sponsorship needed to run the car to its potential, it almost certainly would have. Regardless, this might very well be the greatest racing Falcon New Zealand has produced.

The last person to race Richards' Falcon Sports Sedan was long-time owner Graham Booth, pictured here at Oran Park.

Ford Falcon Sundowner Van Page 39

FORD FALCON SUNDOWNER VAN
The cruisin' Van from Ford

Ford Falcon Sundowner–a van that's right-on when it comes to a combination of performance, strength, room, handling and versatility.

Best of all though is up back–with the easy 'swing-up/drop-down' lift tail-gate. Inside there's a flat steel-ribbed floor that's over 5 feet (1544 mm) wide and is 7 feet (2133 mm) long to the back of the front seats–covered with a rubber floor mat.

Falcon Sundowner gives you bold, good looks, too. Colourful stripes and Sundowner decals. Bonnet scoops. Blacked-out grille with a set of long-range quartz halogen driving lights. For your comfort up front, there are reclining bucket seats and cut pile carpet. A sporty console with armrest and storage bin. Comprehensive instrumentation includes tacho, oil and temperature gauges. Then there's the 'soft-feel' sports steering wheel and column stalk-control for turn/lane change signals, high/low beam, headlight flasher and horn button.

FORD FALCON SUNDOWNER STANDARD EQUIPMENT HIGHLIGHTS

As for Falcon 500 Van but with these upgraded features:

- ★ 4.1 litre crossflow 6-cylinder engine
- ★ 4-speed manual floor shift transmission
- ★ ER70H 14 steel belt radial ply tyres
- ★ Comprehensive instrumentation including tachometer, oil and temperature gauges
- ★ Front centre console with armrest and storage bin
- ★ 'Soft-feel' sports steering wheel
- ★ Passenger compartment cut pile carpet
- ★ Slotted steel sports wheels with chrome trim rings

and with these additional features which are not standard equipment on Falcon 500 Van:

- ★ Long-range quartz halogen driving lights
- ★ Distinctive exterior stripes and 'Sundowner' decals
- ★ Bonnet scoops

and with these Falcon 500 Van features deleted: Bodyside protection mouldings and van side glass

(See page 48 for specifications and available extra cost options.)

Printed January 1979

Top left: Vern Muller has owned and raced this XB hardtop (updated as an XC) since the 1990s.
Above: Philip Macey owns this beautiful replica of the Bryan Byrt XC hardtop raced by Dick Johnson.

The Leo Leonard/Gary Sprague XC Fairmont Ghia pictured during the 1978 B&H 500 at Pukekohe, complete with 302 cu. in V8, automatic transmission, and vinyl roof.

Ford
NO 121

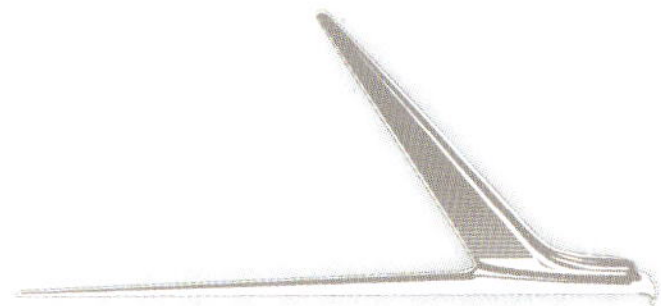

DARRYL'S XC FALCON COBRA

Sometimes legendary cars are created almost by accident, or out of necessity, rather than as a result of any act of careful planning. Such is the case with the limited-edition XC Falcon Cobra, from 1978. The Falcon Cobra is one of the most instantly recognisable of all Aussie muscle cars, but it was created purely for logistical reasons. In essence, it was a marketing exercise intended to offload the last of the outgoing and unwanted Falcon two-door hardtop bodyshells prior to the arrival of the new fourth-generation European styled XD model for 1979. Only 400 were made, and they're now highly sought after by Ford muscle car collectors.

When Ford Australia introduced the new third-generation XA Falcon in 1972, it made available for the first time a two-door hardtop model, whose styling was heavily influenced by the vast and dramatic-looking American Ford Torino. The inclusion of a two-door hardtop was a response to rivals Holden and Chrysler Australia, both of whom had big coupés in their line-up. Holden had added its swoopy Monaro coupé to its HK model range in 1968, which contrasted sharply with the conservative styling of the HK four-door sedan, and it continued the theme with the new HQ range in 1971.

The big coupés offered an exciting and sporty new alternative to the more family-oriented traditional four-door sedans Australian buyers were accustomed to, and while sales were initially strong, as time passed they began to nose-dive as buyers reverted back to the convenience offered by an extra set of doors.

Ford Australia's crisp new XD model, with its European styling heavily influenced by the Granada, was due for release in March 1979, and while sales of the outgoing XC sedan were still going strong, sales of the hardtop had all but ground to a halt. Ford needed to find a way to coax desirability from a body style that was no longer desirable. Enter the Falcon Cobra.

The Falcon Cobra was the brainchild of Edsel Ford II, a great-grandson of Henry Ford. Edsel Ford was doing a stint with Ford Australia at the time, as assistant managing director. The bold Cobra styling was a 1970s take on that of the famed Shelby Mustangs and Shelby Cobra sports cars of the 1960s, but with a '70s twist. The task was handed to Ford designer John Birt, who, a few years earlier, had come up with the attractive two-tone John Goss Special colour scheme released during the XB model run. At the time, the John Goss Special had been a response to slow sales of the coupé model. That strategy had worked in the case of the JGS, and ultimately the XC Cobra strategy would work, too.

Birt laid out the two-tone Cobra colour scheme, consisting of a 'Sno White' base offset with two stripes running down the centre of the car in 'Bold Blue', as well as Bold Blue accents over the wheel arches and sills. According to Ford's Product Planning Manager Ian Vaughan, the Cobras were all painted Bold Blue, with the Sno White applied over the top, and masked off so the blue showed through in the stripes and accents. Bumpers were colour-coded to match the body. A Cobra snake decal was added at the rear of the front mudguards, with a Cobra name decal on the bootlid.

The interior featured tasteful black/blue woollen cloth seats and a sports steering wheel, and each car came with its own individually numbered die-cast plinth, mounted on the glovebox door.

To add to the far-out '70s styling, a vacuum-formed front spoiler was fitted, along with 3-piece rear spoiler, and the whole package rolled on 15 x 7-inch Bathurst Globe alloy wheels.

The Falcon Cobra made quite a statement, and could only ever have been a child of the 1970s, a period when outlandish colour offerings and garish graphics treatments were commonplace. In fact, the concept had already been tested and proven successful in the US with the Mustang Cobra II. If anything, the Mustang Cobra was a more subtle offering.

Ford marketed the Falcon Cobra as a limited-edition model, which is exactly what it was. It was limited to 400 cars, which is how many two-door hardtop bodyshells Ford had yet to sell. But by turning them into a limited edition, Ford would make them more desirable and, it was hoped, they'd all sell. In truth, as sought after as the big Falcon coupé is now, when new it was much less so. Its styling, coupled with high insurance premiums on the high-performance models, kept sales at a crawl, even when the car was first introduced. Barely 500 GT Falcon hardtops had found buyers during the first couple of years of production, and this number had slumped to just 123 by 1975, the second year of the face-lifted XB model. Special editions, such as the John Goss Special, which were cheaper than the GTs, did fairly well, with around 800 units sold. But the hardtop body style was never a big seller. So the XC Cobra was a last-ditch effort to unload the last remaining bodyshells, after which Ford Australia would never venture down the path of producing a Falcon two-door coupé again.

The Cobra Falcons were made available with a number of different engine/transmission options, and each one was individually numbered. Car 001 was a promotional car with a 302 cu. inch V8, and auto. Numbers 002–031 were Bathurst homologation specials, fitted with competition upgrades to benefit the Group C race cars. They were all 351 cu. inch V8 cars with 4-speed top-loader manual gearboxes. The upgrades were actually the modifications homologated for the XC 500 GS hardtop, which first appeared in race competition in September 1977 and which famously won Bathurst that year, with the two Allan Moffat-run machines crossing the finish line in first and second places. These cars differed outwardly from the other Falcon Cobras with their rear-opening bonnet scoop, which was also a part of the

September 1977 XC homologation package. These cars featured items such as transmission oil cooler, strut braces, idler arm brace, electric fan kit, heavy duty radiator, left side anti-tramp rod, reworked rear wheel housings, and long-range fuel tank.

Cars 032-041 were 351 manuals with air-conditioning and power steering. Cars 042-080 were 351 cu. inch manuals with A/C, P/S, and electric windows. Cars 081-200 were 351 cu. inch autos with A/C, P/S, and E/W. Cars 201-300 were 302 cu. inch manuals. Cars 301-360 were 302 cu. inch autos with A/C and P/S. Finally, cars 361-400 were 302 cu. inch autos with A/C, P/S, and E/W.

Of all the Falcon Cobras, the Bathurst Specials (numbers 002-031) are the most desirable, and fetch the highest prices, followed by the 351 cu. inch manual cars. But any Falcon Cobra, regardless of its factory specification, is a very sought-after machine among the Ford faithful.

By 1978, Ford Australia was turning its attention away from producing full-bore high-performance muscle cars as it had done in the late 1960s and early '70s, and consequently the Falcon Cobra in race guise had a slightly underwhelming and extremely brief career. It raced just four times, during the 1978 Championship of Makes, which was the annual endurance series contested at the conclusion of the Australian Touring Car Championship, and which included Bathurst. From the four events, the two factory team cars of Allan Moffat and Colin Bond struggled, and while they possessed enough speed to tackle the now-dominant Holden Torana A9X's head-on, to do so made them fragile, and they invariably failed to finish. However, Bond took an impressive victory in Adelaide, following Bathurst. So while it did not have the glittering career of a GTHO Phase III or a XA GT, the Falcon Cobra, against tough opposition, held up well. But the Cobra was the last in an impressive list of cars built by Ford Australia to be homologated for touring car racing.

Only a handful of Falcon Cobras have made their way to New Zealand, and none would have been sold here when new. The number is thought to be, at most, around ten cars. The car pictured here has been owned for several years by Darryl Lucas. Darryl's car is #121 of the 400 built. Its one of the 120 Cobras fitted with the 351 cu. inch motor and automatic gearbox, along with air conditioning, power steering, and electric windows. Of all the variations, this combination was the most numerous. However, being a 351 cu. inch car, it's also extremely desirable.

Darryl's Cobra was imported into New Zealand in 1985, and in 1987 it was purchased by John Pope. John had lived in Australia in the early 1980s, and stumbled upon a Cobra while there. Smitten, he took a couple of photos of it. Back in New Zealand, upon purchasing this car, he discovered it was the very car he'd photographed in Australia.

Not long after Darryl purchased this car, his wife Shelly bought him a 1:18 scale Biante model of a Falcon Cobra as a gift. Like the real thing, the Biante models were strictly limited edition; each one came with a Certificate of Authenticity and was numbered. Shelly didn't even bother getting into a bidding war. She hit the Buy Now, knowing it was unlikely that another would come on the market. When the model arrived at the Lucas household, they were stunned to discover the model was #121, just like their full-sized version.

Darryl is something of a lifelong Falcon enthusiast, and has owned an XA GT hardtop for decades. He converted it to a classic race car after it arrived as a write-off at the wreckers yard he owned back in the 1980s. He also owned an XB GT sedan with the rarely optioned half-vinyl roof.

OIL
FUEL
TEMP
LAMPS ON
PARK BRAKE

BRIDGESTONE

JD 3698
SUPER DYNAMIC
HANKOOK

SUPER DYNAMIC
HANKOOK

SUPER DYNAMIC
HANKOOK

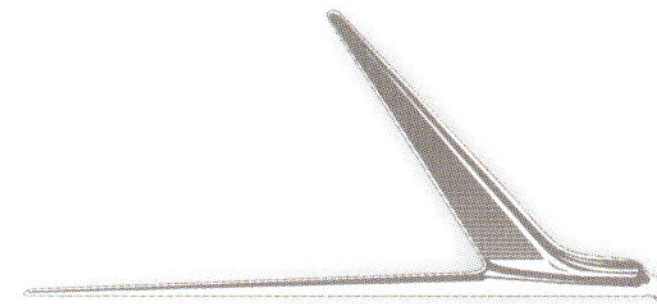

MATT'S
XC

Matt's beautiful 1979 XC Fairmont has been in his family since new. His grandfather ordered it, and ticked all the boxes, including 302 cu. inch V8, black vinyl top, 4-speed auto on the floor, and white interior. He drove it for a number of years, and when he was done, he sold it to his daughter, Matt's mother. She used it as her daily transport, until eventually it was parked in the shed, where it sat for 14 years.

Eventually, Matt took ownership of the Fairmont. He rebuilt the carburettor, got it up and running again, and took it to a paint and panel shop where it was stripped, the rust removed and new metal added. It was then repainted the original factory metallic bronze, and a new vinyl roof was fitted, at the request of Matt's mum. She is still a Falcon enthusiast and, along with her partner, owns an XY.

The period kidney-bean mags were the first choice for those upgrading their Falcons in the 1970s and, with the white letter tyres, they finish the package off perfectly.

Leo Leonard in the XD Fairmont Ghia with which he won the 1981 B&H Endurance Series. The series was open only to New Zealand-assembled cars, and the Fairmont, with its 302 V8 and automatic transmission, was as hot as it got.

Dick Johnson made a handful of guest appearances in Nick Begovic's XD Fairmont Ghia in the B&H Endurance Series.

Above right: The Wayne Wilkinson/Bruce Fowler Fairmont Ghia, pictured at Manfeild during the B&H Endurance Series.
Left: Upon returning to New Zealand from the US in 1990, ex-pat Kiwi Ken Hyndman purchased this Falcon XD to tour the South Island. It had been converted to CNG, and the dealer assured Ken it'd get great gas mileage. He later tried to sell it, but it failed its WOF due to excessive rust, so he gifted it to a friend and it ended its days as a farm hack.

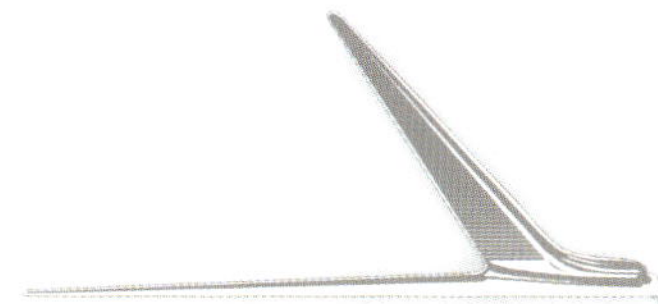

GLENN AND MARIANNE'S

XD FAIRMONT GHIA

The 1980 Ford Fairmont Ghia pictured here is a car I know well. It's been in my family since I was a kid. This was the Holmes' family car. It replaced our 1973 XA Ford Falcon 500 sedan, which itself replaced our Ford Zephyr MkIII. My father is a bit of a Ford nut.

The Fairmont was purchased as a five-year-old car in 1985, having had two previous owners. Although second-hand, it came with a sizable price tag – $20,000: this was in the days before the influx of used Japanese imports drastically reduced the prices of used cars in the New Zealand market.

For us kids, me and my sister Raewyn, the greatest novelty offered by the Fairmont were the electric windows, which were a pretty big deal back in the '80s.

Both my parents, my father Glenn and my mother Marianne, had other cars they used for work, so the Fairmont really served as something we took on holiday. My father is a qualified mechanic, and has always done a lot of the maintenance on his vehicles. He featured a few years ago on a TV3 news segment for the Toyota Hilux he'd owned from new, which had travelled well over 1 million kilometres.

Like the Hilux, the Fairmont was always beautifully maintained. Indeed, it has now travelled 210,000 kilometres, but looks as good as the day it rolled off the assembly line.

Glenn has made a few subtle upgrades over the years, including lowering the suspension slightly, fitting 16-inch wheels, plus a Falcon S-Pack chin spoiler, rear window louvre, sunroof and dark grey paint

along the lower body. He has also fitted an aftermarket extractor and Yella-Terra head. One day, while doing maintenance on the car, Glenn removed one of the rear door cards, which revealed a 1979 stamping, suggesting this to be one of the very earliest XDs built.

The Fairmont reached a stage, at around 15 or 20 years old, when it just didn't have a lot of value. It wasn't yet old enough to be considered a classic, but was no longer new enough to have widespread appeal to regular buyers. We kids had grown up and flown the roost, and it wasn't getting a lot of use. But it had dropped in value so much that my father couldn't see the point in selling it, because it would have been an insult to sell such a pristine car for so little money. So it was kept, and had the cobwebs blown out on the occasional drive to take in a motorsport event or car display.

These days, of course, the values in all classic Australian cars are on the rise, so the decision to hold on to the old Fairmont has proven to be the right one. In addition, my parents enjoy taking it out once a month with their local classic car club. It's inching towards being a forty-year-old car, but its fantastic road manners and crisp, European styling, belie its age.

For this classic Ford, life has gone full circle.

SBR

WARNING
EXIDE

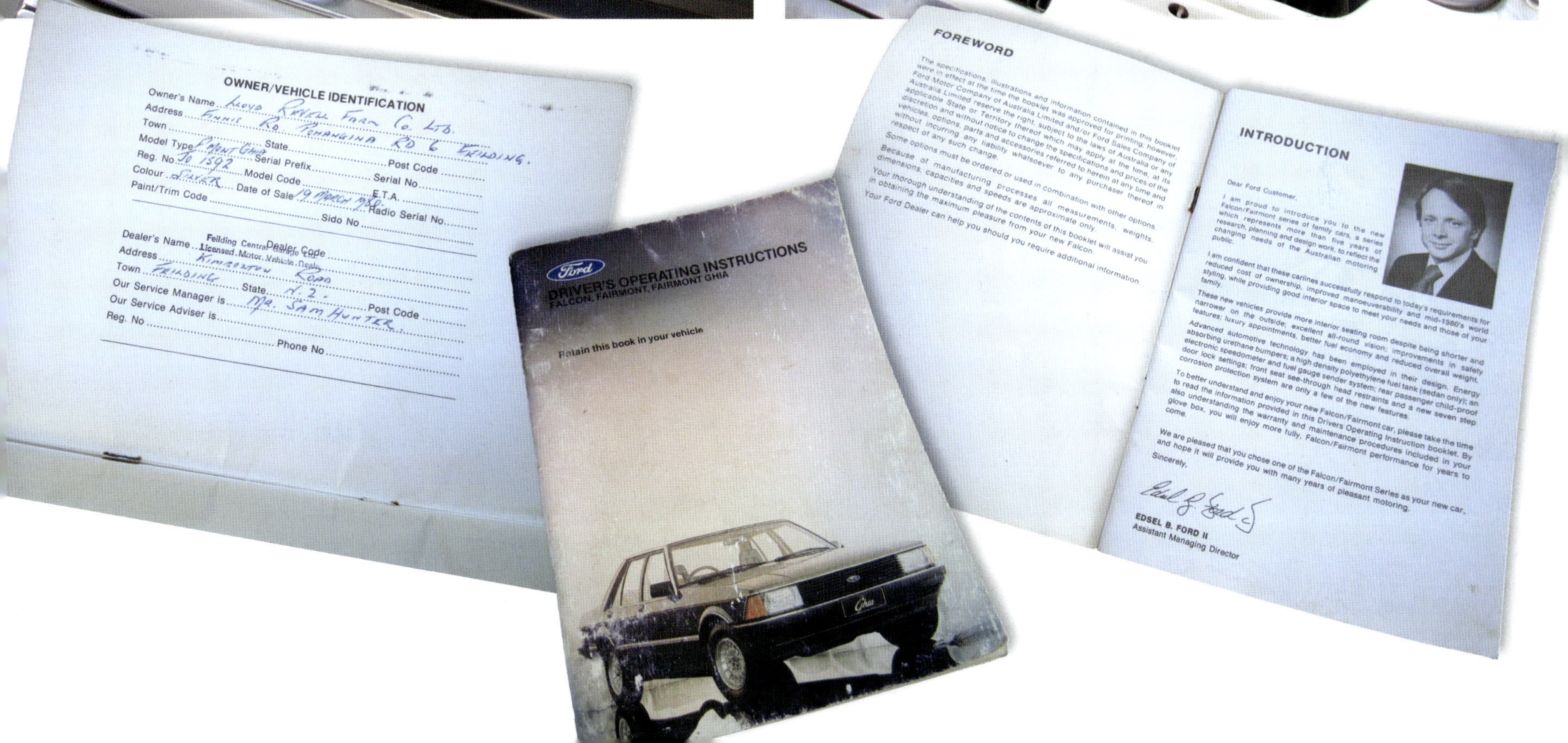
OWNER/VEHICLE IDENTIFICATION
Owner's Name
Address
Town
State
Post Code
Model Type
Serial Prefix
Serial No.
Reg. No.
Model Code
E.T.A.
Colour
Date of Sale
Radio Serial No.
Paint/Trim Code
Sido No.
Dealer's Name
Dealer Code
Feilding Central Garage Ltd
Licensed Motor Vehicle Dealer
Address
Town
State
Post Code
Our Service Manager is
Our Service Adviser is
Reg. No
Phone No.
Ford
DRIVER'S OPERATING INSTRUCTIONS
FALCON, FAIRMONT, FAIRMONT GHIA
Retain this book in your vehicle
FOREWORD
The specifications, illustrations and information contained in this booklet were in effect at the time the booklet was approved for printing; however, Ford Motor Company of Australia Limited and/or Ford Sales Company of Australia Limited reserve the right, subject to the laws of Australia or any applicable State or Territory thereof which may apply at the time, at its discretion and without notice to change the specifications and prices of the vehicles, options, parts and accessories referred to herein at any time and without incurring any liability whatsoever to any purchaser thereof in respect of any such change.
Some options must be ordered or used in combination with other options.
Because of manufacturing processes all measurements, weights, dimensions, capacities and speeds are approximate only.
Your thorough understanding of the contents of this booklet will assist you in obtaining the maximum pleasure from your new Falcon.
Your Ford Dealer can help you should you require additional information.
INTRODUCTION
Dear Ford Customer,
I am proud to introduce you to the new Falcon/Fairmont series of family cars, a series which represents more than five years of research, planning and design work, to reflect the changing needs of the Australian motoring public.
I am confident that these carlines successfully respond to today's requirements for reduced cost of ownership, improved manoeuverability and mid-1980's world styling, while providing good interior space to meet your needs and those of your family.
These new vehicles provide more interior seating room despite being shorter and narrower on the outside; excellent all-round vision; improvements in safety features; luxury appointments, better fuel economy and reduced overall weight.
Advanced automotive technology has been employed in their design. Energy absorbing urethane bumpers; a high density polyethylene fuel tank (sedan only); an electronic speedometer and fuel gauge sender system; rear passenger child-proof door lock settings; front seat see-through head restraints and a new seven step corrosion protection system are only a few of the new features.
To better understand and enjoy your new Falcon/Fairmont car, please take the time to read the information provided in this Drivers Operating Instruction booklet. By also understanding the warranty and maintenance procedures included in your glove box, you will enjoy more fully, Falcon/Fairmont performance for years to come.
We are pleased that you chose one of the Falcon/Fairmont Series as your new car, and hope it will provide you with many years of pleasant motoring.
Sincerely,
EDSEL B. FORD II
Assistant Managing Director

The Steve Midgley/Sid Going Fairmont Ghia tackles the old Pukekohe dog-leg during the B&H Endurance Series.

A once-common sight on New Zealand roads: XD Falcon 'panda' car.

Dean Perkins and the Falcon XD he built for racing from a former police car back in the late 1980s.

Countless old Falcons met their end on speedway bullring ovals as Streetstocks.

ENZED
17
XE2 358ci
Motorcraft
1ST AUTO PARTS
Hankook
NISSAN SPORT 500
ANDERSON &
O'LEARY LTD
PINEPAC
A.Anderson
NZV8
MOSELLE
PANEL & PAINT
TOW

PINEPAC
RACING FALCONS

By the early 1980s, production car racing in New Zealand was witnessing a sea change. The annual Benson & Hedges endurance race at Pukekohe reached a conclusion in 1980, with Rod Coppins and Jerry Clayton winning in a Holden Commodore. The B&H race required cars to be New Zealand assembled, and that they be completely stock. With no option for levelling the playing field, the race was won every year from 1972 to 1978 by a variety of Valiant Charger 770s. Although far from being a performance car, the New Zealand-assembled Charger was still clearly better in a 500-kilometre contest than anything else being locally assembled.

The concept of showroom racing began to lose its meaning when the Charger was still winning the B&H race years after Todd Motors stopped producing it. Plus, endurance events contested in late winter or early spring, as the B&H race was, began to lose their shine with a new generation of race-goers. Therefore, in 1981, the B&H 500 was dropped, and replaced with a three-race endurance series, contested at the three North Island racing circuits: Bay Park, Manfeild, and Pukekohe.

The 1981 B&H endurance series was strongly supported and produced some epic racing. There were four different classes: 0–1350 cc, 1351–1700 cc, 1701–2000 cc and Over 2000 cc. Overseas-assembled vehicles could be entered in the three smaller classes, but in the Over 2000 cc class they had to be New Zealand assembled. David Oxton and Garry Pedersen, driving a Commodore, won the Bay Park race, while Kent Baigent and Graeme Bowkett (in a Toyota Celica) won at Manfeild. Leo Leonard and Gary Sprague, driving a Ford Fairmont Ghia, were first to the chequer in the Pukekohe final, and also won the championship.

The Fairmont Ghia, with its 302 cu. inch V8 and automatic transmission was the toughest locally assembled Ford available. Although not particularly ferocious, it only had to beat Holden's most viable contender, the Commodore SL/E, which was equipped with a 308 cu. inch V8 and auto.

The B&H series proved a big attraction. It also enjoyed some interesting car and driver combinations. The very fast teenager, Paul Radisich, teamed up with his father Frank in a Toyota Trueno. Radisich junior, in only his second racing season, was challenging for the lead at Manfeild when he ran out of fuel. The Ford contingent included the Fairmont Ghia of Nick Begovic, who went to the effort of having Dick Johnson share his car.

For the 1982 series, General Motors New Zealand (GM-NZ) decided to get serious, and took advantage of an untapped loophole in the rules. Although the Over 2000 cc class had to feature New Zealand-assembled cars, manufacturers could build a homologation special that was more focused on performance and racing. There was perhaps an assumption nobody would ever exploit this rule, but GM-NZ duly did. The homologation special was required to be of a mass-produced model, fitted with specific performance components, and built in very small numbers. In the case of the B&H series, that minimum number was just 25 cars. To that end, GM-NZ set about producing 30 Holden Commodore SS models, based on the mass-produced SL/E.

The 308 cu. inch V8 engines fitted to the SS Commodores were all designed and assembled at Peter Brock's HDT Special Vehicles factory in Melbourne. When they emerged from his shop, they were producing 250 horsepower – a big step up from the standard fare. Brock also designed the exhaust system, while the Commodores received further upgrades with a 4-speed M21 manual gearbox and Borg-Warner rear-end. Bilstein shocks, lower springs, larger sway bars, and 15-inch BBS wheels were also used. The regular Commodore SL/E and the Fairmont Ghia were equipped with 14-inch diameter wheels. Among the other racing upgrades were 90-litre fuel tanks replacing the regular 60-litre tank, so the Commodores could run longer between pit stops.

Ford New Zealand had no interest in producing a performance variant of the XD Falcon, and as expected, the Commodores ran away with the 1982 championship. Peter Brock partnered David Oxton in one of two 'factory' SS Commodores (entered by Colin Giltrap) to win the opening encounter at Bay Park, and again at Manfeild, for Round 2. Despite there being at least seven of the new SS Commodores in each race, somehow Leo Leonard and Gary Sprague managed to finish second in the first two races in their Faimont Ghia, which should have been outclassed.

Wayne Wilkinson and Neville Crichton won the final race at Pukekohe, in another SS Commodore, while Brock and Oxton were second, ahead of their team-mates Denny Hulme and Peter Janson in the second factory SS Commodore. Brock and Oxton were declared the champions.

The 1983 B&H series was essentially a repeat of the 1982 one, with results being dominated by the SS Commodores. There was no factory team as there had been the previous year, but it didn't matter. Ray Smith purchased the Commodore raced in 1982 by Wilkinson and Crichton, and he brought in Hulme to share driving duties. And the pair emerged as champions. Furthermore, most of the Ford teams had walked away.

Building on the growing popularity for touring car racing in New Zealand, a new series was added to the motorsport calendar for the 1983/84 season, sponsored by ANZ bank. Whereas the B&H endurance series took up its regular spot in late 1983, the ANZ series would follow the annual summer racing international programme kicking off in December 1983, and continuing on into early 1984. The rules for the ANZ series were based on those of the B&H series, but with several freedoms introduced to avoid one-make domination. The freedoms extended to wheels and tyres, engine components and exhausts,

and suspension parts. Significantly, the imported Falcon, with its 351 cu. inch V8 and 4-speed gearbox, was now eligible.

One of the Fairmont Ghia teams contesting the B&H series was that of brothers Bruce and Wayne Anderson. They'd purchased the car raced by Nick Begovic and Dick Johnson during the 1981 series. The Andersons were already well versed in production racing, having run a brace of MkIV Zodiacs and a Falcon GTHO Phase III. When the new rules for 1983/84 were announced, they set about building a new XE Falcon 351.

Other than its obvious cosmetic changes, the biggest difference between the new XE Falcon and the outgoing XD was the introduction of coil-spring rear suspension and Watts-link for the first time. The Falcon had used leaf springs in the back since its inception, and coils were intended to produce a better-quality ride in the road cars. But they should have equated to better performance in the race cars as well. To achieve this, however, took time. Dick Johnson began racing an XE in late 1982, and struggled initially get make it as fast as his old XD. But inevitably, it did happen.

The Andersons started with a brand-new bare bodyshell purchased in Australia and flown to New Zealand. The shell was acid dipped to get as much weight out as possible, then seam-welded. Construction began with the fitment of an aluminium roll cage, while Koni shocks were fitted front and rear. Front springs rates were a mightily stiff 1300 pounds to carry the hefty 351 cu. inch Cleveland V8, while rear

spring rates were 280 pounds. Competition suspension and steering bushes replaced the stock items, while a big 1.25-inch sway bar was mounted in the front.

The 351 engine was race prepared and equipped with aluminium heads, and produced 400 horsepower. It was backed by a Borg-Warner Super T-10 4-speed gearbox. Under the regulations, the team had to retain the 8-inch Borg-Warner rear-end with no floating axles, which would prove to be the big Falcon's Achilles heel. Minimum racing weight for the Falcon under the rules was 1420 kilograms.

Although not strictly legal, the Falcon was fitted with a big 3-piece rear spoiler, the same type being used on the Group C racing Falcons in Australia. It was finished in an attractive shade of yellow/orange, similar to the old factory McLarens of Bruce McLaren and Denny Hulme that dominated the Can-Am sports car series in the United States in the late 1960s and early '70s. Sponsorship came from the Anderson family business, Anderson O'Leary Ltd Pinepac.

The Falcon was driven predominantly in the 1983/84 ANZ series by Bruce. A second Falcon, run as a Ford Dealers entry, was raced by Gary Sprague. Sprague's was an older XD model. Both Falcons were highly competitive throughout the championship. Sprague went on to win in his XD, while Anderson finished third. The rules provided plenty of grey areas, and the Anderson car was prevented from even starting the opening round at Bay Park when it was found to be fitted with an aluminium radiator, incorrect carburettor, incorrect differential, and illegal suspension modifications.

For the 1984/85 season, New Zealand went full-on into using FIA Group A touring car regulations, which were already in widespread use throughout Europe and the United Kingdom, in the European Touring Car Championship and the British Saloon Car Championship. Furthermore, Australia was about to switch to Group A for the 1985 Australian Touring Car Championship.

In addition, a major new international event was to be contested on the streets of Wellington in January 1985, followed by another race at Pukekohe a week later. The two-race series was sponsored by Nissan Sport and attracting a strong entry of top Group A cars from around the world. Each race was over a distance of 500 kilometres.

Under Group A regulations, the Pinepac Falcon could no longer use aluminium heads. It did, however, receive a set of larger, 16-inch, Simmons wheels. Bruce once again contested the ANZ series, now called the New Zealand Touring Car Championship. Impressively, he won! But it was at the 1985 Wellington Street Race that the big Falcon really shone on an international stage.

The great Pinepac Falcon, at the 1985 Wellington Street Race, where, against a field of top international Group A cars, Dick Johnson blew everyone's doors off by comfortably taking pole position. Race day saw a different outcome, unfortunately.

Andrew Anderson's wild Central Muscle Cars XE race car sits alongside his road car, which, for this photo, was fitted with the wheels from his family's original Pinepac Falcon.

Although Bruce shared the Falcon with younger brother Wayne in the long-distance races, for the Nissan series, he would co-drive with Australian Ford racing hero, Dick Johnson. Anderson already had a small involvement with Johnson through Ron Richards, who worked at DJR. The Andersons had gone to Richards looking to coax more power out of the 351 engine. As Australia was switching to Group A for 1985, Johnson had to forgo his successful Falcons for the third-generation Fox-body Mustang. Ford Australia had no interest in homologating the Falcon for Group A, so it simply couldn't race in its own country. And with the latest variation on the third-generation Mustang yet to be homologated, Johnson didn't have a car to drive if he wished to contest the New Zealand events.

Johnson was already well acquainted with the Pinepac Falcon. He'd taken up an invitation to drive the car with Bruce at a Pukekohe endurance event in 1984, where the wheels literally fell off. During practice, the fragile rear axles broke, and the wheels parted company. The team was waiting back in the pits, unaware of the dramas, and wondering where their hot-shot had got to. Eventually he walked back in, and proclaimed, 'Mate, she was like trying to land a 747 in a tail wind!'

Despite the Pukekohe antics, Johnson took up another invitation from the Andersons to share the Falcon at the 1985 Wellington Street Race. His great Holden rival, Peter Brock, would be there, with the first-ever Holden Dealer Team Group A Commodore. Larry Perkins would share this car with Brock. Other entries included one of the wickedly fast Rover Vitesses being raced in Europe by Tom Walkinshaw Racing. Walkinshaw would share this car with New Zealander Ron Dickson. Another of the fast European cars, a turbocharged Volvo 240T was entered by Mark Petch. This car had been competing in the Belgian Touring Car Championship. It'd be driven by its Belgian driver Michel Delcourt, and Kiwi Robbie Francevic. In addition, there was a fleet of BMW 635CSi's, including a former Schnitzer car for Kent Baigent and Neil Lowe, a former JPS car for Neville Crichton and Wayne Wilkinson, plus Frank Sytner's example that had been competing in the British Saloon Car Championship. Sytner would share with John Morton.

Nobody really fancied the changes of the Pinepac Falcon, particularly as it was the biggest car in the field, and the Wellington street circuit was tight and slow. But straight out of the box, Johnson was punching in times his rivals simply couldn't match. These were the fastest Group A machines in the world, and a locally built Falcon was outrunning them.

The pace of the big Falcon continued into qualifying, where Johnson blitzed the field and qualified on pole position. His best time was over a second faster than the next car's, the BMW of Crichton and Wilkinson. Johnson's pole time was 1 min 32.49. He was the only driver to record a lap under 1 min 33 – and what's more, he repeated the feat two more times.

But while the Falcon was blindingly fast, there were already concerns within the Pinepac team as to its durability. On a regular, smooth racing circuit, the Falcon's fragile rear-end had proven to be its weakest link. On the bumpy Wellington waterfront, it was getting a hammering like never before. The Detroit Locker was unlocking itself on the tight turns, then locking up again with a thud, placing extra strain on the spindly axles.

Come the race, and Johnson blasted away from the standing start to build a healthy lead early on. After three laps, the chasing pack was over 300 metres in arrears. Johnson soon caught lapped traffic, which held him up through the tighter sections. But his lead continued to grow. However, after twelve laps, Johnson coasted into the pits. The wheel bearings in the rear-end had started working loose, and he sensed failure was imminent. A long pit stop followed while repairs were made, then he was sent out again. Anderson also got a drive but, ultimately, the Falcon retired.

The following weekend at Pukekohe, with Johnson at the wheel, the axles broke, and the rear of the Falcon was launched in the air. It was a stark reminder of the volatile nature of racing a vehicle the manufacturer didn't want to race.

Following the Nissan series, Johnson went home and prepared his Mustangs for the ATCC. Although it displayed potential, the Falcon would only serve as a temporary measure in Group A. While it never wanted for horsepower, the regulations required it retain much of the stock rear componentry. Any manufacturer involved in homologating a car for Group A would simply ensure it was offered with a driveline able to cope with the rigours of racing. Had Ford Australia wanted to race the Falcon in Group A, it would have gifted it a much beefier rear-end, and bigger brakes. But it didn't want to race it. So the Andersons were always on borrowed time. And sure enough, by late 1985, they were racing the first of what would ultimately be two Pinepac Group A Mustangs.

The Falcon continued to be raced. Warren Midgley, the Andersons' cousin, campaigned it for a short time while Bruce was sidelined with a broken collarbone. It was then sold to Ian Midgley, another relative of the Andersons. Sadly, it was later stolen, following a restoration, and has never been recovered. And gone with it was an important and significant piece of New Zealand motor-racing history.

Although the car is gone, the Pinepac Falcon story lives on through Bruce and his son Andrew. After Bruce had stepped away from Group A racing, he started competing in classic events in a 1969 Mustang fastback which he owns to this day. Andrew, very early in his racing

career, drove the Mustang in Sports & GT events. Eventually, he got involved in touring car racing, competing in both the NZV8 series and later the V8 Supertourers. He also did a stint at Bathurst in 2009, having got a late call to drive the Greg Murphy Racing VZ Commodore in the Supercars development series race.

Back in 1983 when construction of the XE Pinepac Falcon began, Andrew was just 11 years old. His primary school was right next to the family business, so he would spend time in the shop each morning before heading to school. Watching the Falcon come to life left a lasting impression on him. Having Dick Johnson drive the car sealed the deal.

Despite the long list of cars Andrew has raced over the years, the XE Falcon just stuck in his mind. So in 2015, he started looking for a road car to purchase. After looking at a few options, he struck on a factory 302 V8 car with power steering and 4-wheel disc brakes. A previous owner had installed a 351. It needed some love, and was just the kind of project he was after. He drove it for a few months; then, in February 2016, he parked it in the Pinepac race shed, stripped it back and carried out a full restoration. He had it repainted in the same colour as the old family Pinepac racer, and had completed the restoration by January 2018.

Meanwhile, Andrew's uncle Wayne had purchased an XE for his son to drive. In August 2017, when the motor needed replacing, Wayne bought another XE (a Fairmont) and removed its engine. Andrew had planned to buy the donor Fairmont, but then, following a family meeting, Bruce elected to have Wayne's business AV8 Motorsport build it into a Central Muscle Cars Group 1 race car for Andrew to drive. Of course, the paint scheme would be a tribute to the old Pinepac Group A car – using the colour known internally as Pinepac Toby Yellow.

Since this car first hit the track in 2018, a few have mistakenly assumed it to be the original Pinepac Group A. But they couldn't be more wrong. Whereas the Group A Falcon was essentially a modified road car, Central Muscle Cars are a mix of modern technology and monstrous horsepower packed into a classic bodyshell.

The motor is a 358 cu. inch Roush Yates Motorsport unit built for NASCAR road courses. It packs an epic 875 horsepower, more than double that of the old Group A car. To cope with all that power, a bulletproof driveline is required. A NASCAR rated 4-speed Andrews gearbox transmits all those horses through an aluminium driveshaft to a Ford 9-inch differential with MS Fabrication cambered hubs. Global West A-arms and lower control arms are used in the front, with strengthened factory 4-link and Watts-link in the rear. Koni double-adjustable coil-overs are fitted front and rear, with King Springs. Brakes are by Brembo. The Group A Pinepac XE ran on 15-inch diameter BBS wheels with polished rims and gold centres, as does the tribute XE, albeit much larger ones – at 17 x 9.5.

Understandably, the new Pinepac Falcon is significantly faster than the old one. At more than double the horsepower, with bigger wheels and brakes, and with a much more rigid chassis, you'd expect that. But the new version is truly a weapon. Since it debuted, it's been a leading contender in Central Muscle Cars, which is already packed with enormous firepower. CMC was invited to run with the Touring Car Masters support category at Bathurst in 2019, where these epic machines really got to stretch their legs. Current V8 Supercars top-out at just under 300 km/h down Conrod Straight. When Andrew drove the Greg Murphy Racing VZ V8 Supercar at Bathurst in 2009, he was clocking 289 km/h. That was completely tapped-out. And yet, in the Pinepac XE, he was hitting 308 km/h. With far less downforce and a more primitive chassis and suspension design, that's bound to get your attention.

For Andrew and Bruce, catching up with Dick Johnson during that Bathurst visit and having him look over the epic tribute Pinepac XE was a highlight. Dick happily approved having his name on the tribute, which pays respect to the original.

NZV8 magazine did a feature on the Pinepac Falcon tribute back in 2019. The location for the photos was, of course, the Pinepac yard in Whenuapai. So much racing history has come from this place. One of the photos included both the racing tribute XE and Andrew's own XE street car. The Midgley family, who owned the Group A after the Andersons, still have the original wheels, which are all that remain of the car. They very kindly let Andrew fit them to his street car for the photo shoot. Sadly, the original Group A XE couldn't be here for the photo shoot. But this is surely the next best thing.

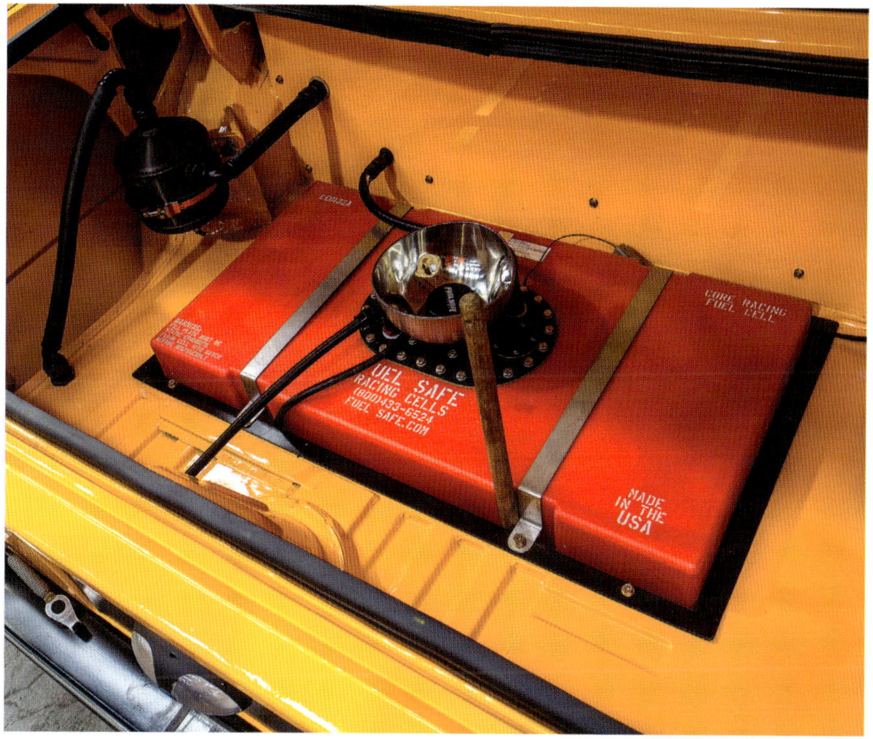

Andrew (left) with his father Bruce. Bruce began racing this 1969 Mustang Sportsroof in the late 1980s, and these days competes in the Central Muscle Cars category with Andrew.

Justin Butcher's Central Muscle Cars XE Falcon, adorned in the Australian racing legend Dick Johnson's famous racing Greens-Tuf colours that his cars wore throughout 1983 and '84.

Top left: Ray Peterson's blown Chrysler Hemi-powered XF Falcon Wild Bunch drag car, which later became the roof-chopped Dave and Chad Tulley machine.
Top right: Duelling Falcons in the Super-6 series at Manfeild.
Middle right: The very first DJR Falcon V8 Supercar, pictured during the 1992 Wellington Street Race. Dick Johnson and John Bowe qualified fourth but failed to finish.
Above: The Mal Rose/Mark Larkham Falcon EB2 pictured at the 1992 Wellington Street Race. These days Larkham is a popular V8 Supercars TV presenter.

CRC
10

GWS289

WITH $3000 WORTH OF EXTRAS
AT NO EXTRA COST
THIS FALCON IS A CLASSIC.
Factory Fitted CFC-Free Air Conditioning.
Classic Decal & Pin Stripe.
Prestige Paint Colours Available
FALCON CLASSIC
Sports Alloy Wheels.
P205 - R15 Tyres.
THE LIMITED EDITION FALCON CLASSIC.
Ford
HAVE YOU DRIVEN A FORD...LATELY?

Ford
CEK338

During the Pukekohe round of the 1996 Mobil series, the EF Falcon of Dick Johnson prepares to leave pit lane. The 1996 Mobil series was the last time the Wellington Street Race was held.

FALCO
YES WE CAN
XQ2958
John Dyke Ford

JARED'S

XHII XR8 UTE

This is Jared's stunning XR8 ute. It is perhaps the ultimate wolf in sheep's clothing: fast, light and unexpected – not like a ute at all! It's also totally mint.

It's a one-family vehicle. Jared's father-in-law Dave ordered it new in November 1998. At the time, Jared had only been with his now-wife for three months. Dave had a list of options he wanted in the ute, including air conditioning and electric windows. But when he placed his order, it was already being built, and the cost to change the specification was too high. He could have waited and ordered one of the new AU models which were about to be launched, but he didn't want one of those. So he stuck with the XHII package as supplied.

To say this has been a pampered vehicle is an understatement. It has travelled 130,000 kilometres since new, and still has its original exhaust, windscreen, and clutch. Its completely unmodified, and incredibly original.

In 2019, after having owned the XR8 for two decades, Dave passed it on to Jared, knowing it'd continue to be cared for, and would remain in the family. Since taking ownership, Jarod has enjoyed cruising the XR8 on his days off and hopes that, in time, his young daughter will eventually sit her driver's licence in it.

FUEL INJECTION
Ford
BEWARE OF FAN
DANGER

XO2958

FALCON'S LANDING
XO2958

FALCON
XR8
XR8
XQ2958
John Dyke Ford

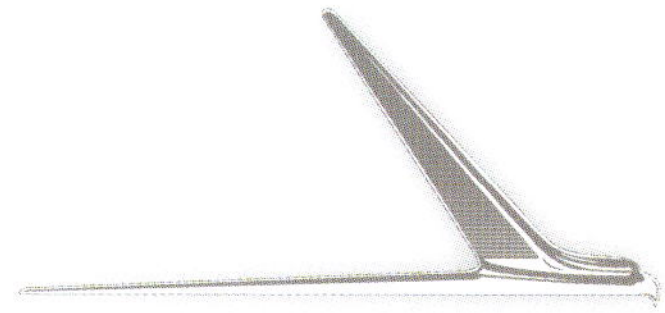

BLAIR'S
AU TS50

The 'GT' nametag will forever be synonymous with the Falcon. GT is purely a branding exercise used by Ford to label its sporty models. It is most closely linked to the high-profile American Fords, such as the GT40 and Shelby GT350. But its use has had a good workout. It first appeared in the mid-1960s, and was then applied to virtually every sporty Ford throughout the remainder of the decade, and into the '70s.

Consider the Fairlane GT and GTA, Torino GT, Ranchero GT, Shelby GT500, Mustang GT and GTA. It was even attached to the British Fords - for example, the Cortina. Ford Australia built a special MkI Cortina in 1965 to contest the Bathurst endurance event. Back then, Bathurst was contested over 500 miles. The purpose-built Cortina was thus named the GT500.

The GT badge first appeared on an Australian Falcon with the arrival of the XR model in late 1966. The Falcon GT remained in the Ford arsenal for ten years, until the XC was launched in July 1976. It continued through three phases of the high-end GTHO, but Ford decided to retire it when it released the XC, as it continued its shift away from high-performance vehicles.

In 1991, in a joint venture with Tickford Vehicle Engineering, Ford Australia finally revived the Falcon GT. It was designed as a high-end grand tourer, and featured the plush interior appointments of the Fairmont Ghia. The Falcon GT was produced until 1997, before it was again mothballed. In 2002, however, with the release of the BA range, the GT was reinstated as the ultimate performance Falcon, and remained so until the very last Falcon, the FG-X, was launched in 2014.

But in between the ongoing appearance and disappearance of the GT label, the Falcon still boasted some significant sporty themes. When Ford dropped the GT option with the release of the XC, it promoted the GS as its top performance model. Then, in 1978, it launched the Falcon Cobra. The Cobra was partially a marketing exercise to rid the company of its outgoing two-door hardtop bodyshells, with the new XD model launch looming. But the Cobra was also a genuine performance car, and sported numerous components aimed at making it a winner on the track. The XD and XE range offered the S-Pack and, more significantly, the ESP. And with the launch of the EB in 1992 came the first of the XR6 and XR8 models.

In 1991, Tickford Vehicle Engineering was established as the Ford rival to Holden Special Vehicles, producing high-performance variants of the regular models, plus other engineering projects a large-scale manufacturer wasn't capable of achieving. Tickford created the XR range of sporty Fords and in 1998, shortly after the release of the AU Falcon, the T-Series range. Three models were produced in the T-Series range: the TE50, TS50, and TL50. The TL50 was based on the long wheelbase Fairlane model.

The TE50, TS50, and TL50 all featured handsome bodykits which greatly improved the polarising design of the AU Falcon. The TE50 and TL50 featured a 270-horsepower (200-kW) 5-litre V8 based on the Windsor motor, whereas the TS50 was fitted with aluminium cylinder heads, and packed a full 300 horsepower (220 kW). By comparison, the Falcon XR8 produced 248 horsepower (185 kW). Each model was fitted with high-performance suspension and brake packages, and while the TE50 wore 17-inch diameter wheels, both the TS50 and TL50 went to 18s. The TE50 used a 5-speed manual transmission with the option of a 4-speed automatic, while the TS50 and TL50 were only available with the auto.

All three models were produced in very small numbers, which virtually guaranteed their status as a future classic: TE50 – 104, TS50 – 54, and TL50 – 43.

With the release of the AUII in 2000 came the second-generation of T-Series models, which gained various upgrades. All three were now produced with the 300-horsepower engine, and the TS50 could now be optioned with the 5-speed manual.

Again, tiny numbers were produced: TE50 – 141, TS50 – 46, and TL50 – 12.

With the AUIII in 2001 came the third and final generation of the T-Series. The third-generation featured a more aggressive bodykit, and horsepower was increased to match, with a 5.6 litre stroker version of the Windsor V8, producing 340 horsepower (250 kW). Ultimately, these were the most powerful normally aspirated Falcons until the eventual release of the Boss 290 GT Falcon.

Performance increased, as did build numbers: TE50 – 204, TS50 – 224, and TL50 – 3. Nevertheless, these were still very low production numbers, and the cars are highly sought after among Ford performance enthusiasts.

Ultimately, the T-Series should probably have been named the Falcon GT, but Tickford was aiming to build on its brand name. Regardless, these were truly some of the best Fords to emerge from the late-1990s/early-2000s era. They are rarely seen in Australia, and even less so in New Zealand.

For performance enthusiasts, the AUIII-based TS50 is the ultimate of the full T-Series range. And that's what features here. Blair's TS50 is one of the 224 examples built on the AUIII platform, and one of just 50 that made their way to New Zealand. Of course, it's got the manual gearbox, just as any enthusiast's performance car should have. Blair purchased it in 2013, and it has travelled just 42,000 kilometres from new. Its part of a wider collection of neat Aussie Fords owned by Blair and his wife Helena, which also includes an XA Falcon GT hardtop, and Helena's mild custom XP wagon (see Chapter 2).

TS50

ENHANCED
BEWARE
OF FAN

Top left: Paul Radisich 'The Rat', in his Dick Johnson Racing AU Falcon during the Pukekohe round of the 2001 V8 Supercars Championship.
Above: Author Steve Holmes' Falcon XR6 during a project car mission in 2007.

Liam MacDonald's flame-spitting Falcon New Zealand Touring Car pictured at Pukekohe.

Falcon
MEANXB

APY655

LNE375

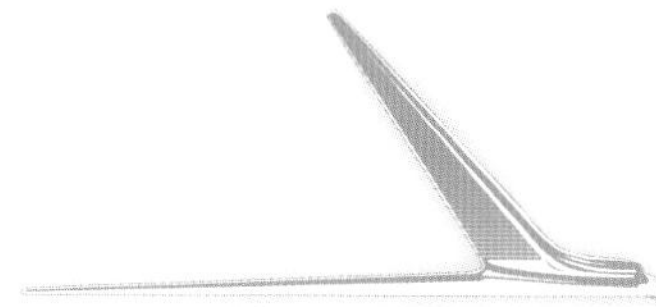

KELVIN

Like many Falcon enthusiasts, Kelvin is a serial collector, not content with just one example. His fleet of Falcons covers almost all the bases, from classic to modern, track to street.

The oldest in the collection is a beautiful 1973 XB two-door hardtop. It's fitted with a 351 cu. inch backed by a 4-speed Top Loader, and 9-inch differential. It's been lowered 2 inches and wears a strut-brace as well as front and rear sway bars to help it carve through the corners. Brakes are Brembo 4-piston in the front, and Willwoods in the back. The 18-inch wheels are offset by redline tyres, and its heft is eased a little with the fitment of a carbon fibre bonnet.

Kelvin is currently building a 400 cu. inch Cleveland with Nitrous-Oxide for the XB. The silver wagon is an EA with EB GT front. It's truly a wolf in sheep's clothing, powered by a Tickford XR6 engine and 5-speed gearbox with Tickford suspension.

The AU was built as a track car, for endurance racing. It's an XR6 VCT Cobra, with variable-cam timing. It has been lightened, rolls on 18-inch wheels, and sports Tickford suspension.

The newest car in the fleet is a beautiful 2003 BA XR6 Turbo. It's a Harrod Motorsport build, runs a 4-speed sequential gearbox, and rolls on 18-inch wheels.

Essentially, Kelvin has a Falcon for every occasion.

MARK'S
FG FPV PURSUIT

Since its inception in the 1920s, Australians have embraced the 'ute' with enthusiasm. The first Australian ute (short for 'utility vehicle') was considered to have been produced by Ford, having been prompted by a farmer's wife requesting a vehicle that could be used to attend church on Sunday and take pigs to market on Monday. Thus, the utility was created, melding the front half of a car with the back half of a truck.

Ford Detroit introduced the Ranchero in late 1956. It was based on the short wheelbase platform of the full-sized Ford models, and combined the front half of a sedan body with integrated custom cargo bed. The Ranchero was referred to as a coupé utility, and spawned an entirely new market segment. It was, in essence, a stylish work vehicle, and one that could appease both car and pick-up buyers. Indeed, many Rancheros never saw work duties at all.

In 1959, Chevrolet countered with the El Camino. It was essentially a two-seat Impala with a swooping rear roofline and semi-enclosed tray.

The second-generation Ranchero was based on the compact Falcon platform, and when the Falcon was picked up by Ford Australia, the Ranchero debuted in Australia, albeit called simply the Falcon Utility.

Although marketed purely as a work vehicle, the Falcon ute's style and versatility saw it gain a strong foothold in the Australian market – much more so than the Ranchero in the US. Indeed, a ute remained an important part of the Falcon range until the model's demise in 2016. And while most utes were purchased purely as work-hacks, many became much-loved family transport. Furthermore, spotting an opening, both Ford and rival Holden started offering optional upgrades, including V8 engines, style packages and the like, from the late 1960s.

Perhaps the most extreme version of the Falcon ute was the Ford Performance Vehicles (FPV) Pursuit. This was, in essence, a ute variant of a Falcon GT.

This particular 2014 FG FPV Pursuit is owned by Mark. It is incredibly rare; Ford/FPV only produced 120 of them, and only three came to New Zealand. It's fitted with the supercharged 5-litre Coyote V8, producing 315 kW (approximately 420 horsepower), and backed by a 6-speed auto. Because of its rarity, Mark has kept it factory stock, although he has lowered the suspension and fitted an aftermarket exhaust.

The Falcon GT pictured here with the Pursuit was also Mark's. It's gone now, but this wild ute remains. It's certain the farmer's wife who wrote to Ford in 1932 requesting a combined church-runner and pig-hauler vehicle had no idea what her letter would eventually spawn.

FEB110

GT F

HAN230

Left: Craig Baird's United Videos Falcon V8 Super Tourer pictured at Manfeild.
Above: Kiwi V8 Supercar driver Scott McLaughlin won the first of three V8 Supercar Championships in 2018, aboard his DJR Team Penske FG-X Falcon. 2018 was also the last year the Falcon was eligible to compete in the championship, thus the great model was retired on a high.

Ford
FPV
GT
XC GXL

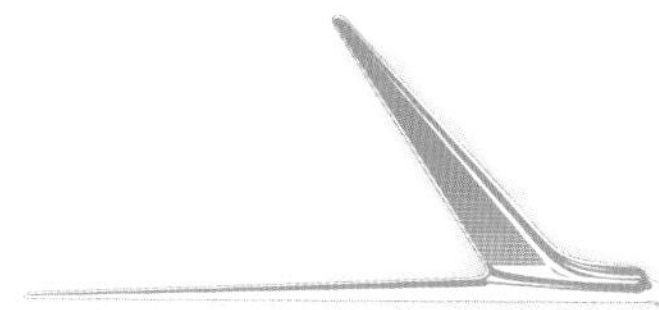

BRYAN

When he was a young bloke, Bryan worked hard and built a very successful trucking company that afforded him the ability to indulge in his hobbies. For years he competed in various forms of motorsport – predominantly speedway. He won the New Zealand Street Stock Championship aboard a Valiant Charger, and eventually worked his way into Sprintcars, the true centrepiece of the sport in New Zealand. Strapped into a 650 kilogram projectile with 900 horsepower skating sideways around a dirt oval lined with a concrete wall is probably the ultimate rush.

But away from the high-octane thrills and spills, Bryan also enjoyed a more sedate hobby: restoring classic Mack trucks. And it was the truck restoration work that evolved into his incredible Falcon collection.

The cars in Bryan's collection are only the rarest of the rare. In the case of the newer examples, they're all ultra-low-volume high-performance models. The classics, however, while relatively rare even when they were new, are even more so today due to low survival rates of early Falcons.

Like most collections, Bryan's just naturally grew. Each new arrival highlighted a gap and so another car was acquired.

The oldest car in the collection is a stunning yellow XA hardtop. It's easy to forget, given it was the high-end GTs that were winning on the track, that a large proportion of these big birds were ordered with a 250 cu. inch inline-6. Throughout the 1970s and '80s, many 6-cylinder Falcons were fitted with V8s. But not this one ... and that's

why it appealed to Bryan. It's a factory 6-cylinder, it's still fitted with its original engine, and Bryan is just the second owner. When it came into his possession, he had it fully restored, but everything about it is as it was when it rolled off the assembly line in 1973, except for the period-correct slot mags. These were installed by its original owner.

This is a numbers-matching car, and backing the 170-horsepower 6 cylinder is the original 4-speed gearbox.

Parked next to the XA hardtop is an equally impressive XB GT sedan. Again, this is a numbers-matching car, and is also completely original, although the big 351 has been worked over.

Bryan also has an XC. This is a cream with vinyl roof 1978 GXL, and Bryan had it imported from Australia. It's incredibly original, although it has been repainted once in the correct colour. Ford Australia dropped the GT name plate when it launched the XC model, and other than the brief run of Cobra hardtops in 1978, the GXL was effectively the hot performer. It sported the GT bonnet with the twin scoops, and various other performance features, and was available as either a sedan or a hardtop. Bryan's GXL is ultra-rare, being one of just four with the options it has.

Fourth-generation 'Blackwood' Falcons are represented in the form of an impressive XE ESP. The ESP was available in both the XD and XE, and while based on the Fairmont Ghia, was the ultimate performer of the fourth-generation models before the V8 engine was dropped from the line-up. This is one of just 360 XE ESP's built with the 302 engine, and one of 58 with these colour and specification options.

Fifth-generation Falcons are well represented in Bryan's collection. First up is a 1992 EB GT. Ford Australia resurrected the GT badge to celebrate the 25th anniversary of the first GT Falcon. As well as having lavish performance upgrades, these cars also sported an eye-popping bodykit that truly justifies its important name plate. Only 265 were produced, and only 15 came to New Zealand. This is number 5.

Fittingly, right next to the 25th-anniversary GT is a 30th-anniversary

GT. It was this model that *Wheels* magazine took to Germany to thrash up and down an Autobahn. Back in 1971, *Wheels* magazine writer Mel Nichols was testing a new GTHO Phase III with photographer Uwe Kuessner. The pair had slept late on the final morning of the multi-day test and had to get the Falcon back to Ford in Melbourne, some 320 kilometres from where they'd been staying in Albury. With only two hours to reach their destination, Nichols opened up the big 351 and kept it open, with the throttle hard on the floor. The Phase III swept up to 145 mph (233 kph), and while sitting at that speed, Kuessner climbed into the back seat and took a photo of the speedo. The pair reached Melbourne on time, and when the road test went to print, the photo had been doctored to show only 104 mph ... just to save any aggravation that might follow.

The Kuessner photo is widely considered the most famous in Australian motoring history. This was the reason *Wheels* magazine decided to recreate it with an EL GT. Given no Australian road could be used, Nichols and Kuessner instead took a car to Germany, and as expected, hit their target.

Only 272 EL GT's were produced, and Bryan's is one of just 20 to be sent to New Zealand.

A third fifth-generation Falcon in the collection is owned by a good friend of Bryan's, who lives in the US. This is truly a rare sight, even in its homeland Australia. It's an Alan Jones Special. Alan Jones is best known as being only the second Australian driver behind Jack Brabham to win the Formula 1 World Championship. By the mid-1980s, he was carving out a new career away from F1. Initially, he raced predominantly in sports cars, but increasingly became involved in touring car racing in his homeland. After several special guest drives, he found a more permanent spot with Tony Longhurst's team before joining Glenn Seton in the early days of the new V8 Supercars formula.

In 1996, Jones and Seton's main sponsor Phillip Morris International set out on its own, with Kiwi brothers Ross and Jimmy Stone joining

to build a fleet of Falcons as part of a two-car team. With cigarette sponsorship being banished from the sport, the team was named Pack Leader Racing, and Kiwi driver Paul Radisich, who was doing great things in the British Touring Car Championship at the time, guest-drove Jones' second Falcon at the very last Wellington Street Race in 1996.

Because Jones was such a big name in Australian motorsport, and given his association with the Ford brand, it was inevitable that a small batch of high-performance Falcons bearing his name would eventually be produced. These are incredibly rare cars. It's believed only two of them reside in New Zealand.

Continuing the GT anniversary theme is a 2007 BF II 40th anniversary. Two hundred of these were produced, and only 25 of those were originally imported to New Zealand when new. Bryan had been looking for one of these for a while, but couldn't find a local one he liked, so imported this one from Australia. He is just the second owner, and it's like new.

Surprisingly, this collection started with Bryan's stunning white 2014 FPV GT F. The 'F' in the name refers to this being the 'Final' Falcon GT model ever to be produced. Ford Australia already knew in 2014 that the end was coming for Falcon production, and decided to retire the celebrated GT name plate with a special run of 500 cars. Fifty of these came to New Zealand, and Bryan purchased this one brand new.

Parked next to the GT F is another rarity in the form of a silver 2003 TS50. This was the last in the three generations of the T-Series Tickford Falcons based on the AU range; this being the AUIII. Three variations were produced: the TE50, TS50, and TL50. This, the TS50, was the ultimate performer, producing 250 kW (340 horsepower). Only 224 AUIII TS50s were produced, and this is one of the 50 sent to New Zealand.

Finally, although it's not a Falcon, Bryan's magnificent 2018 Mustang 'Bullitt' special edition neatly finishes off the set. With Falcon production ending in 2016, the Mustang took its place as the halo model in the Australian and New Zealand markets for blue oval loyalists. Furthermore, the Mustang replaced the Falcon in the Supercars Championship from 2019. As is typical of Bryan's epic collection, the Mustang is an anniversary model, celebrating 50 years since Steve McQueen's blockbuster Bullitt hit the silver screens in 1968. Only 50 of these cars reached New Zealand's shores, and Bryan purchased this one brand new.

Apart from the fact they're both two-door Fords that seat four people and have the power delivered to the rear wheels, the XA Falcon and Bullitt Mustang are totally different cars. They're separated by 45 years, and several technological generations. But the pair book-end this incredible collection perfectly.

GT
ELGT 18

Ford
FPV
W.A.
GT 40 168

GTF
351 FPV

Tickford
T
250 TS

"Have you driven a Ford... lately?"
FORD
FALCON
Ford Motor Company of Australia

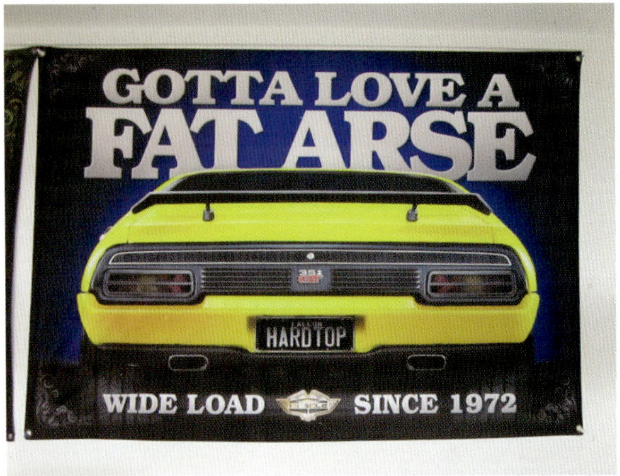
GOTTA LOVE A
FAT ARSE
HARDTOP
WIDE LOAD
SINCE 1972

New Zealand
Classic Car
TWO FAVOURITE FALCONS
Taranaki
Sprint Car
Championship
2011-2012

GT

Ford
FPV
GT
250 TS

DICK JOHNSON
17
TRU-BLU
QUICKIES
Pump and Polish
FILLER UP
GET YOUR ROD POLISHED HERE!
SATISFACTION GUARANTEED!
"Have you driven a Ford... lately?"

08
09.
10
11
12
13
14
Taranaki Sprint Car Championship 2011-2012

MUSTANG
MCQUEEN

PHOTO CREDITS

Anderson Family Collection: page 159
Murray Beatson: page 11, 12 left, 23 top left and bottom right, 29 bottom left, 33 top left, 57 top left and bottom right, 75 top left, 94 top left, 118 centre and bottom left, 124 top, 138 bottom left, 147 top right, 148 top left, 155 bottom, 166 top left and top right, 168 centre left, bottom left and bottom right, 178 right, 179 left and bottom right, 184, 185 top left
Graeme Bennett: pages 130, 169
James Berryman: pages 60–69
Blair: pages 24–27, 174–177
Gary Bogaart Collection: pages 120–123
Graham Booth: page 137 bottom
Malcolm Cambridge: page 178 top left
Allan Cameron: page 119 top right
Euan Cameron: page 179 top right, 198
Ross Cammick: page 75 bottom, 119 bottom right, 124 bottom right, 125, 136, 139, 154 top
Warwick Clayton: page 12 right, 32, 35, 52, 56, 82 bottom right, 85 centre and bottom left, 87 bottom left, 97 top and centre left, and right, 104 top, 105 bottom, 118 bottom right, 119 top left and centre right, 124 bottom left, 132, 133, 137 top, 138 top left, 147 bottom left and bottom right, 148 top right, 154 bottom centre and right, 155 top right, 160 left, 160 right, 166 bottom left and bottom right, 167, 168 top left, 185 top right and bottom left, 196, 200
Wayne Collins: pages 16–21
Martin Derbyshire: page 135 left
Perry Drury: pages 48,49
Mike Feisst: pages 22 bottom left, 46, 84 left
Milan Fistonic: page 97 top right
Rod Grimwood Collection: page 135 right
James Hancox: pages 70–73
Peter Hanna Collection: page 51 top and bottom, 82 top left and bottom left, 84 right, 85 top left

Hayden: pages 106–112
Ken Hyndman: page 148 bottom
Peter Jacobs: pages 88–93
Jared: pages 170–173
Tony Johnson: pages 95, 96
John Judson: page 104 bottom
Kelvin: page 180
Philip Macey: page 138 bottom right
Mark: page 183
Marty: pages 126–129
Matt: page 146
McCaughan Family Collection: page 58 top
Rick McLaren: pages 76–79
O. J. McLean: page 154 bottom left
Phil Noble: page 97 bottom left
NZV8 magazine: pages 140–145, 156, 162–165
Parkes Family Collection: page 75 top right
Grant Paterson at Motografix: page 57 top right and bottom left
John Pope Collection: pages 98–103
Allan Porter: pages 4, 30, 59, 74, 94 top right and bottom right
Stewart Quertier: pages 28, 29 top, 94 bottom left
Gerard Richards: pages 23 top right and bottom left, 33 top right and bottom right, 34 top and bottom right, 155 top left, 166 centre left
Rob: pages 36–45
George Sheweiry: page 87 top left
Garry Simkin: page 33 bottom left
Martin Smith: pages 160 centre, 161
Sprague Family Collection: pages 31 bottom right, 82 top right, 85 top right
Ross Tebbs: page 58 bottom right
Bruce Thompson: page 85 bottom right
Steve Twist: page 55, 80, 94 centre left, 118 top right
Neville Webb: page 185 bottom right

JOHN JONES STEEL
A.Knight
5
Gulf
Go Further
Milwaukee
HAGLEY
THE ROCK FM
CRC
S.Richards
Further
J.McIntyre
47
Go Further
amcor
NORWOOD
Castrol
EDGE
EDGE
R&B

ACKNOWLEDGEMENTS

It's hard to know where to start with this one. I have so many people to thank, because so many people contributed to this book.

Thanks to all the people who opened up their garages and shared their stories and the many, many people who have kindly donated their photos. These are personal photos people have kept for years, showing the many varied roles these cars have played in the lives of Kiwis. These amazing people have been listed in the photo credits section.

In addition, I'd like to thank Peter Bryan and Mark Fiddyment for supplying me with an amazing selection of old car sales brochures. In my opinion, it was the addition of these brochures that really helped lift this book to another level.

Thank you to all the people who provided leads and multiple photos, including Todd Wyllie, Warwick Clayton, Ross Cammick, Murray Beatson, Steve Twist, Rod Grimwood, Pete Jacobs, and Graeme Scown.

Thanks to my parents Marianne and Glenn for ensuring I became a petrolhead!

Thanks to Bill Honeybone and the team at David Bateman Publishing for producing another winner.

Last but not least, thank you to my wonderful but long-suffering wife Helen, who puts up with me travelling away for long periods, and then locking myself in my office when I'm actually home, as I do the final big push to get these books finished. I really appreciate you!

PENSKE
Truck Rental
12
HOGS
HOGS
MAN
MAN
PIRTEK
PIRTEK
PPG
PPG
PENSKE
Truck Rental
PENSKE

ABOUT THE AUTHOR

Steve Holmes is a New Zealand motoring writer and motorsport historian who has written mostly motorsport history articles for various New Zealand and Australian car magazines for over twenty years.

Although he has a broad range of interest in all things motoring related, in particular, Steve is a motor racing enthusiast. His passions are for saloon car racing, both New Zealand and international, from the 1960s through 1980s, as well as the American Can-Am sports car series, and Formula 5000.

Steve has written more than fifteen books, both for New Zealand and for international publishers, and his motorsport history books in particular have been widely praised for their research and depth of detail. In recent years he has performed commentary duties at historic racing events.

In 2011, Steve created a motorsport history website and discussion forum called The Roaring Season (www.theroaringseason), to which he and site members contribute various stories and photographs, covering not only motorsport history from around the world, but also modern-day historic racing events. Many of the photos appearing in this book were unearthed through The Roaring Season.

Steve has owned numerous hobby cars over the years, mostly of the classic American V8 variety. He recently imported a historic racing car from the United States, for which he plans a cosmetic restoration, before finally taking part in historic racing events.

DUNLOP
JOHN JONES STEEL
A.Knight
Gulf
Go Further
Milwaukee
THE ROCK FM
HAGLEY
CRC
S.Richards
556
NORWOOD
V8 SuperTourers NEW ZEALAND